Texts in Computing

Volume 24

Domain-Specific Languages of Mathematics

Texts in Computing Series Editor
Ian Mackie   mackie@lix.polytechnique.fr

Domain-Specific Languages of Mathematics

Patrik Jansson

Cezar Ionescu

Jean-Philippe Bernardy

ISBN 978-1-84890-388-3

College Publications
Scientific Director: Dov Gabbay
Managing Director: Jane Spurr

http://www.collegepublications.co.uk

Cover produced by Laraine Welch

Contents

About this book

Software engineering involves modelling very different domains (e.g., business processes, typesetting, natural language, etc.) as software systems. The main idea of this book is that this kind of modelling is also important when tackling classical mathematics. In particular, it is useful to introduce abstract datatypes to represent mathematical objects, to specify the mathematical operations performed on these objects, to pay attention to the ambiguities of mathematical notation and understand when they express overloading, overriding, or other forms of generic programming. We shall emphasise the dividing line between syntax (what mathematical expressions look like) and semantics (what they mean). This emphasis leads us to naturally organise the software abstractions that we develop in the form of domain-specific languages, and we will see how each mathematical theory gives rise to one or more such languages, and appreciate that many important theorems establish "translations" between them.

Mathematical objects are immutable, and, as such, functional programming languages are a very good fit for describing them. We shall use Haskell as our main vehicle, but only at a basic level, and we shall introduce the elements of the language as they are needed. The mathematical topics treated have been chosen either because we expect all students to be familiar with them (for example, limits of sequences, continuous functions, derivatives) or because they can be useful in many applications (e.g., Laplace transforms, linear algebra).

0.1 Origins

This book started out as lecture notes aimed at covering the lectures and exercises of the BSc-level course "Domain-Specific Languages of Mathematics" (at Chalmers University of Technology and University of Gothenburg). The immediate aim of the book is to improve the mathematical education of computer scientists and the computer science education of mathematicians. We

1

believe the book can be the starting point for far-reaching changes, leading to a restructuring of the mathematical training of engineers in particular, but perhaps also for mathematicians themselves.

Computer science, viewed as a mathematical discipline, has certain features that set it apart from mainstream mathematics. It places much more emphasis on syntax, tends to prefer formal proofs to informal ones, and views logic as a tool rather than (just) as an object of study. It has long been advocated, both by mathematicians [Wells, 1995, Kraft, 2004] and computer scientists [Gries and Schneider, 1995, Boute, 2009], that the computer science perspective could be valuable in general mathematical education. Until today, as far as we can judge, this perspective has been convincingly demonstrated (at least since the classical textbook of Gries and Schneider [1993]) only in the field of discrete mathematics. In fact, this demonstration has been so successful, that we see discrete mathematics courses being taken over by computer science departments. This is a quite unsatisfactory state of affairs, for at least two reasons.

First, any benefits of the computer science perspective remain within the computer science department and the synergy with the wider mathematical landscape is lost. The mathematics department also misses the opportunity to see more in computer science than just a provider of tools for numerical computations. Considering the increasing dependence of mathematics on software, this can be a considerable loss.

Second, computer science (and other) students are exposed to two quite different approaches to teaching mathematics. For many of them, the formal, tool-oriented style of the discrete mathematics course is easier to follow than the traditional mathematical style. Moreover, because discrete mathematics tends to be immediately useful to them, the added difficulty of continuous mathematics makes it even less palatable. As a result, their mathematical competence tends to suffer in areas such as real and complex analysis, or linear algebra.

This is a serious problem, because this lack of competence tends to infect the design of the entire curriculum.

We propose that a focus on *domain-specific languages* (DSLs) can be used to repair this unsatisfactory state of affairs. In computer science, a DSL "is a computer language specialized to a particular application domain" (Wikipedia, 2021-12-27), and building DSLs is increasingly becoming a standard industry practice. Empirical studies show that DSLs lead to fundamental increases in productivity, above alternative modelling approaches such as UML [Tolvanen, 2011]. Moreover, building DSLs also offers the opportunity for interdisciplinary activity and can assist in reaching a shared understanding of intuitive or vague notions. This is supported by our experience: an example is the work done at Chalmers in cooperation with the Potsdam Institute for Climate Impact Research in the context of Global Systems Science, Lincke et al. [2009], Ionescu and Jansson [2011, 2012], Botta et al. [2017a,b, 2018].

Thus, a course on designing and implementing DSLs can be an important addition to an engineering curriculum. Our key idea is to apply the DSL approach to a rich source of domains and applications: mathematics. Indeed, mathematics offers countless examples of DSLs: in this book we cover types and complex arithmetics (Chapter 1), sets and logic (Chapter 2), functions and derivatives (Chapter 3), algebras and morphisms (Chapter 4), power series (Chapter 5), differential equations (Chapter 6), linear algebra (Chapter 7), Laplace transforms (Chapter 8), probability theory (Chapter 9). The idea that the various branches of mathematics are in fact DSLs embedded in the "general purpose language" of set theory was (even if not expressed in these words) the driving idea of the Bourbaki project[1] which exerted an enormous influence on present day mathematics.

Hence, the topic of this book is *DSLs of Mathematics*. It presents classical mathematical topics in a way which builds on the experience of discrete mathematics: giving specifications of the concepts introduced, paying attention to syntax and types, and so on. For the mathematics students, the style of this book will be more formal than usual, as least from a linguistic perspective. The increased formality is justified by the need to implement (parts of) the languages. We provide a wide range of applications of the DSLs introduced, so that the new concepts can be seen "in action" as soon as possible. For the computer science students, one aspect is to bring the "computer-aided learning" present in feedback from the compiler from programming to also help in mathematics.

In our view a course based on this textbook should have two major learning outcomes. First, the students should be able to design and implement a DSL in a new domain. Second, they should be able to handle new mathematical areas using the computer science perspective. (For the detailed learning outcomes, see Fig. A.1 in Appendix A.)

To achieve these objectives, the book consists of a sequence of case studies in which a mathematical area is first presented, followed by a careful analysis that reveals the domain elements needed to build a language for that domain. The DSL is first used informally, in order to ensure that it is sufficient to account for intended applications (for example, solving equations, or specifying a certain kind of mathematical object). It is in this step that the computer science perspective proves valuable for improving the reader's understanding of

---

[1] The Bourbaki group is the pseudonym of a group of mathematicians publishing a series of textbooks in modern pure mathematics, starting in the 1930:s. See Wikipedia, 2021-12-27.

the mathematical area. The DSL is then implemented in Haskell. Finally, limitations of the DSL are assessed and the possibility for further improvements discussed. More about the course is presented in Appendix A.

0.2   Who should read this book?

The book is recommended for Haskell developers who are learning maths and would like to use Haskell to create concrete models out of abstract maths concepts to improve their understanding.

The book explores the connection between mathematical structures and Type-Driven Development [Brady, 2016] of Haskell programs. If you enjoyed "The Haskell Road to Logic, Maths and Programming" [Doets and van Eijck, 2004], you might also enjoy this book.

It is also a book for the mathematically interested who wants to explore functional programming and domain-specific languages. The book helps put into perspective the domains of Mathematics and Functional Programming and shows how Computer Science and Mathematics can be usefully studied together.

We expect the reader to have knowledge corresponding to a few first-year mathematics and computer science courses, preferably including functional programming. But we will review all mathematics using our methodology, and keep to a restricted subset of Haskell (no "advanced" features are required).

Working knowledge of functional programming is helpful, but it should be possible to pick up quite a bit of Haskell along the way.

The book is not primarily a collection of ready-made code-snippets, but rather uses Haskell as a "tool for learning". Even code which cannot actually run can still be useful in the quest to debug the mathematical understanding of the reader (through scope-checking and type-checking) when trying to encode different mathematical concepts in Haskell.

It would be an interesting endeavour to port the code from Haskell to a language with an even stronger type system, like Agda [Norell, 2009], Idris [Brady, 2016], or Lean [de Moura et al., 2015]. The authors would welcome contributions in this direction.

0.3 Notation and code convention

The book is a collection of literate programs: that is, it consists of text interspersed with code fragments (in Haskell). The source code of the book (including in particular all the Haskell code) is available on GitHub in the repository `https://github.com/DSLsofMath/DSLsofMath`.

Our code snippets are typeset using `lhs2tex` [Hinze and Löh, 2020], to hit a compromise between fidelity to the Haskell source and maximize readability from the point of view of someone used to conventional mathematical notation. For example, function composition is typically represented as a circle in mathematics texts. When typesetting, a suitable circle glyph can be obtained in various ways, depending on the typesetting system: `&#8728` in HTML, `\circ` in TEX, or by the RING OPERATOR Unicode codepoint (U+2218), which appears ideal for the purpose. This codepoint can also be used in Haskell (recent implementations allow any sequence of codepoints from the Unicode SYMBOL class). However, the Haskell Prelude uses instead the infix operator . (period), as a crude ASCII approximation, possibly chosen for its availability and the ease with which it can be typed. In this book, as a compromise, we use the period in our source code, but our typesetting tool renders it as a circle glyph. If, when looking at typeset pages, any doubt should remain regarding to the form of the Haskell source, we urge the reader to consult the github repository.

The reader is encouraged to experiment with the examples to get a feeling for how they work. But instead of cutting and pasting code from the PDF, please use the source code in the repository to avoid confusing error messages from indentation and Unicode symbols. A more radical, but perhaps more instructive alternative would be to recreate all the Haskell examples from scratch.

Each chapter contains exercises to help the reader practice the concepts just taught. Sometimes the chapter text contains short, inlined questions, like "Exercise 1.8: what does function composition do to a sequence?". In such cases there is some more explanation in the exercises section at the end of the chapter, and the exercise number is a link to the correct place in the document.

In several places the book contains an indented quote of a definition or paragraph from a mathematical textbook, followed by detailed analysis of that quote. The aim is to improve the reader's skills in understanding, modelling, and implementing mathematical text.

Some more advanced material that can be skipped is marked with a star (*).

0.4 Acknowledgments

The support from Chalmers Quality Funding 2015 (Dnr C 2014-1712, based on Swedish Higher Education Authority evaluation results) is gratefully acknowledged. Thanks also to Roger Johansson (as Head of Programme in CSE) and Peter Ljunglöf (as Vice Head of the CSE Department for BSc and MSc education) who provided continued financial support when the national political winds changed.

Thanks to Daniel Heurlin who provided many helpful comments during his work as a student research assistant in 2017.

This work was partially supported by the projects GRACeFUL (grant agreement No 640954) and CoeGSS (grant agreement No 676547), which have received funding from the European Union's Horizon 2020 research and innovation programme.

Bernardy is supported by the Swedish Research Council, via grant 2014-39, funding the Centre for Linguistic Theory and Studies in Probability.

Thanks also to Jane Spurr at College Publications for helping in publishing the book [Jansson et al., 2022].

The authors also wish to thank several anonymous reviewers and students who have contributed with many suggestions for improvements.

Chapter 1

Types, Functions, and DSLs for Expressions

In this chapter we exemplify our method by applying it to the domain of types and functions first, and complex numbers second, which we assume most readers will already be familiar with. While doing this we also introduce some of the Haskell concepts needed later.

We will implement certain concepts in Haskell and the code for this chapter is placed in a module called *DSLsofMath.W01* that starts here. It is strongly recommended to try out the different examples, to play with the code and to make different edits and tests in order to reach a deeper understanding. As mentioned earlier, the code is freely available on GitHub for this purpose.

```haskell
module DSLsofMath.W01 where
import Numeric.Natural (Natural)
import Data.Ratio (Ratio, (%))
```

These lines constitute the module header which usually starts a Haskell file. We will not go into details of the module header syntax here but the purpose is to "name" the module itself (here *DSLsofMath.W01*) and to **import** (bring into scope) definitions from other modules. As an example, the last line imports types for rational numbers and the infix operator $(\%)$ used to construct ratios ($1 \% 7$ is Haskell notation for $\frac{1}{7}$, etc.).

1.1 Types of data and functions

Dividing up the world (or problem domain) into values of different types is one of the guiding principles of this book. We will see that keeping track of types can guide the development of theories, languages, programs and proofs. We follow a Type-Driven Development style of programming.

1.1.1 What is a type?

As mentioned in the introduction, we emphasise the dividing line between *syntax* (what mathematical expressions look like) and *semantics* (what they mean).

As an example of DSL we start with *type expressions* — first in mathematics and then in Haskell. To a first approximation one can think of types as sets. The type of truth values, *False* and *True*, is often called *Bool* or just $\mathbb{B}$. Thus the name (syntax) is $\mathbb{B}$ and the semantics (meaning) is the two-element set $\{\textit{False}, \textit{True}\}$. Similarly, we have the type $\mathbb{N}$ whose semantics is the infinite set of natural numbers $\{0, 1, 2, \dots\}$. Other common mathematical types are $\mathbb{Z}$ of integers, $\mathbb{Q}$ of rationals, and $\mathbb{R}$ of real numbers. The judgment $e : t$ states that the expression e has type t. For example $\textit{False} : \textit{Bool}$, $2 : \mathbb{N}$, and $\sqrt{2} : \mathbb{R}$. In Haskell, double colon (::) is used for the typing judgment, but we often use just single colon (:) in the mathematical text.

So far the syntax for types is trivial — just names. Every time, the semantic is a set. But we can also combine these names to form more complex types.

Pairs and tuple types For a pair, like $(\textit{False}, 2)$, the type is written $(\textit{Bool}, \mathbb{N})$ in Haskell. In general, for any types A and B we write (A, B) for the type of pairs. In mathematics, the type (or set) of pairs is usually called the *Cartesian product* and is written using an infix cross: $A \times B$. We will sometimes use this notation as well. The semantics of $\textit{Bool} \times \textit{Bool} = (\textit{Bool}, \textit{Bool})$ is the set $\{(F, F), (F, T), (T, F), (T, T)\}$ where we shorten *False* to F and *True* to T for readability. We can also form expressions and types for triples, four-tuples, etc. and nest them freely: $((17, \textit{True}), (\sqrt{2}, \texttt{"hi"}, 38))$ has type $((\mathbb{N}, \mathbb{B}), (\mathbb{R}, \textit{String}, \mathbb{N}))$.

List types If we have a collection of values of the same type we can collect them in a list. Examples include $[1, 2, 3]$ of type $[\mathbb{N}]$ and $[(\texttt{"x"}, 17), (\texttt{"y"}, 38)]$ of type $[(\textit{String}, \mathbb{N})]$. The semantics of the type $[\textit{Bool}]$ is the infinite set

$$\{[\,], [F], [T], [F, F], [F, T], [T, F], [T, T], \dots\}$$

1.1.2 Functions and their types

For our purposes the most important construction is the function type. For any two type expressions A and B we can form the function type $A \rightarrow B$. Its semantics is the set of "functions from A to B" (formally: functions from the semantics of A to the semantics of B). As an example, the semantics of $\mathbb{B} \rightarrow \mathbb{B}$ is a set of four functions: $\{const\ False, id, \neg, const\ True\}$ where $\neg : \mathbb{B} \rightarrow \mathbb{B}$ is Boolean negation. The function type construction is very powerful, and can be used to model a wide range of concepts in mathematics (and the real world). But to clarify the notion it is also important to note what is *not* a function.

Pure and impure functions   Many programming languages provide so-called "functions" which are actually not functions at all, but rather procedures: computations depending on some hidden state or exhibiting some other effect. A typical example is $rand(N)$ which returns a random number in the range $1..N$. Treating such an "impure function" as a mathematical "pure" function quickly leads to confusing results. For example, we know that any pure function f will satisfy the property: if $x == y$ then $f(x) == f(y)$. As a special case we certainly want $f(x) == f(x)$ for every x. But with $rand(\cdot)$ this does not hold: $rand(6) == rand(6)$ will only be true occasionally (if we happen to draw the same random number twice). Fortunately, in mathematics and in Haskell all functions are pure.

Because function types are really important, we immediately introduce a few basic building blocks to construct functions. They are as useful for functions as zero and one are for numbers.

Identity function   For each type A there is an *identity function* $id_A : A \rightarrow A$. In Haskell all of these functions are defined once and for all as follows:

$$id :: a \rightarrow a$$
$$id\ x = x$$

In Haskell, a type name starting with a lowercase letter is a *type variable*. When a type variable (here a) is used in a type signature it is implicitly quantified (bound) as if preceded by "for any type a". This use of type variables is called "parametric polymorphism" and the compiler gives more help when implementing functions with such types. We have seen one example use of the identity function already, as one of the four functions from *Bool* to *Bool*. That instance of *id* has type *Bool* $\rightarrow$ *Bool*.

Constant functions   Another building block for functions is *const*. Its type mentions two type variables, and it is a function of two arguments:

$$const :: a \rightarrow b \rightarrow a$$
$$const\ x\ \_ = x$$

The underscore (_) is here used instead of a variable name (like y) which is not needed on the right hand side (RHS) of the equality sign. Above we saw the instance *const False* : *Bool* $\rightarrow$ *Bool* where a and b are both *Bool*. Note that this is an example of *partially applied* function: *const* by itself expects two arguments, thus *const False* still expects one argument.

The term "arity" is used to describe how many arguments a function has. An n-argument function has arity n. For small n special names are often used: binary means arity 2 (like $(+)$), unary means arity 1 (like *negate*) and nullary means arity 0, thus not a function at all, just any regular value (like "hi!").

Higher-order functions   We can also construct functions which manipulate functions. They are called *higher-order* functions and as a first example we present *flip* which flips the order the two arguments of a binary operator.

$$flip :: (a \rightarrow b \rightarrow c) \rightarrow (b \rightarrow a \rightarrow c)$$
$$flip\ op\ x\ y = op\ y\ x$$

As an example *flip* $(-)$ 4 10 == 10 − 4 == 6 and *flip const x y* == *const y x* == *y*.

Lambda expressions   It is possible to create values of a function type without naming them, so-called "anonymous functions". The syntax is $\lambda x \rightarrow b$, where b is any expression. For example, the identity function can be written $\lambda x \rightarrow x$, and the constant function could also be defined as *const* $= \lambda x\ \_ \rightarrow x$. The ASCII syntax uses backslash to start the lambda expression, but we render it as a Greek lower case lambda.

Function composition   The composition of two functions f and g, written $f \circ g$ and sometimes pronounced "f after g" can be defined as follows:

$$f \circ g = \lambda x \rightarrow f\ (g\ x)$$

As an exercise it is good to experiment a bit with these building blocks to see how they fit together and what types their combinations have.

The type of function composition is perhaps best illustrated by a diagram (see Fig. 1.1) with types as nodes and functions (arrows) as directed edges. In

$$
\begin{array}{ccc}
a & & \\
g\downarrow\ \ \nearrow\!\!\!\!\!\searrow f\circ g & & \\
b\ \xrightarrow{\ f\ }\ c & &
\end{array}
$$

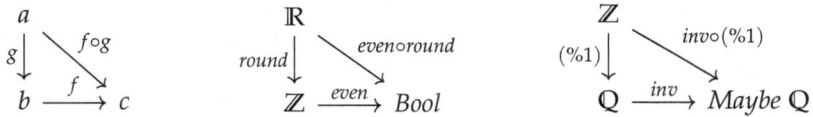

Figure 1.1: Function composition diagrams: in general, and two examples

Haskell we get the following type:

$$(\circ) :: (b \to c) \to (a \to b) \to (a \to c)$$

which may take a while to get used to.

In the figure we use "operator sections": $(\%1) :: \mathbb{Z} \to \mathbb{Q}$ is the function that embeds an integer n as the ratio $\frac{n}{1}$. Other convenient examples include $(+1) ::$ $\mathbb{Z} \to \mathbb{Z}$ for the "add one" function, and $(2*)$ for the "double" function.

1.1.3 Partial and total functions

There are some differences between functions in the usual mathematical sense, and Haskell functions. Some Haskell "functions" are not defined for all inputs — they are *partial* functions. Simple examples include *head* :: $[a] \to a$ which is not defined for the empty list and $(1/) :: \mathbb{R} \to \mathbb{R}$ which is not defined for zero. A proper mathematical function is said to be *total*: it is defined for all its inputs. In Haskell totality can be compromised by omitting cases (like *head*), by raising exceptions (like division) or by non-termination (like *inf* = $1 + inf$).

There are two ways of turning a partial function into a total function. One can limit the type of the inputs (the domain) to avoid the inputs where the function is undefined (or non-terminating, etc.), or extend the type of the output (the range) to represent "default" or "exceptional" values explicitly.

As an example, $\sqrt{\cdot}$, the square root function, is partial if considered as a function from $\mathbb{R}$ to $\mathbb{R}$. It can be made total if the domain is restricted to $\mathbb{R}_{\geq 0}$, or if the range is extended to complex numbers. In most programming languages the range is extended in another way. The type is *Double* $\to$ *Double* and $\sqrt{-1}$ returns the value *NaN* : *Double* (Not a Number). Similarly, $(1/) :: Double \to$ *Double* returns *Infinity* :: *Double* when given zero as an input. Thus *Double* is a mix of (many, but not all) rational numbers and some special quantities like *NaN* and *Infinity*.

Often the type *Maybe a* with values *Nothing* and *Just a* (for all $x :: a$) is used as the target of functions which would otherwise be partial: any undefined input is mapped to *Nothing*. The definition of *Maybe* is given in full in Section 1.2.

There are also mathematical functions which cannot be implemented at all (uncomputable functions). We will only briefly encounter such a case in Section 2.2.1.

Partial functions with finite domain   Later on (in Section 1.7.3), we will use partial functions for looking up values in an environment. Here we prepare this by presenting a minimal DSL for partial functions with a finite domain. The type *Env v s* will be the *syntax* for the type of partial functions from *v* to *s*, and defined as follows:

> **type** *Env v s* = $[(v,s)]$

As an example value of this type we can take:

> *env1* :: *Env String Int*
> *env1* = $[(\text{"x"}, 17), (\text{"y"}, 38)]$

The intended meaning is that "x" is mapped to 17, etc. The semantic domain is the set of partial functions, and, as discussed above, we represent those as the Haskell type $v \rightarrow Maybe\ s$.

Our evaluation function, *evalEnv*, maps the syntax to the semantics, and as such has the following type:

> *evalEnv* :: *Eq v* $\Rightarrow$ *Env v s* $\rightarrow$ $(v \rightarrow Maybe\ s)$

This type signature deserves some more explanation. The first part (*Eq v* $\Rightarrow$) is a constraint which says that the function works, not for *all* types *v*, but only for those who support a Boolean equality check ((==) :: $v \rightarrow v \rightarrow Bool$). The next part of the type signature (*Env v s*) shows the type of the first argument (*env*) to the function *evalEnv*. The final part of the type, $(v \rightarrow Maybe\ s))$, shows that *evalEnv env* is also a function, now taking a *v* and maybe returning an *s*.

The implementation proceeds by searching for the first occurrence of *x* in the list of pairs (v, s) such that *x* == *v*, and return *Just s* if one is found, and *Nothing* otherwise.

> *evalEnv vss x* = *findFst vss*
>   **where** *findFst* $((v,s) : vss)$ | *x* == *v*   = *Just s*
> | *otherwise* = *findFst vss*
>   *findFst* $[\ ]$ = *Nothing*

Another equivalent definition is *evalEnv* = *flip lookup*, where *lookup* is defined in the Haskell Prelude with the following type:

> *lookup* :: *Eq a* $\Rightarrow$ *a* $\rightarrow$ $[(a,b)]$ $\rightarrow$ *Maybe b*

1.1.4   Variable names as type hints

In mathematical texts there are often conventions about the names used for variables of certain types. Typical examples include f, g for functions, i, j, k for natural numbers, x, y for real numbers and z, w for complex numbers.

The absence of explicit types in mathematical texts can sometimes lead to confusing formulations. Here, and in many places in later chapters, we will analyse a quote from a mathematical textbook. You do *not* need to understand the mathematics behind at this point (we only get to the Laplace transform in Chapter 8). For example, a standard text on differential equations by Edwards, Penney, and Calvis [2008] contains at page 266 the following remark:

> The differentiation operator D can be viewed as a transformation which, when applied to the function $f(t)$, yields the new function $D\{f(t)\} = f'(t)$. The Laplace transformation $\mathcal{L}$ involves the operation of integration and yields the new function $\mathcal{L}\{f(t)\} = F(s)$ of a new independent variable s.

This is meant to introduce a distinction between "operators", such as differentiation, which take functions to functions of the same type, and "transforms", such as the Laplace transform, which take functions to functions of a new type. To the logician or the computer scientist, the way of phrasing this difference in the quoted text sounds strange: surely the *name* of the independent variable does not matter; the Laplace transformation could very well return a function of the "old" variable t. We can understand that the name of the variable is used to carry semantic meaning about its type (this is also common in functional programming, for example with the conventional use of a plural "s" suffix, as in the name xs, to denote a list of values.).

Rather than relying on lexical or syntactical conventions in the variable names, we prefer to explicitly use different types. When there are several interpretations of the same type, we can define a type synonym for each interpretation. In the example of the Laplace transform, this leads to

$$\mathcal{L} : (T \to \mathbb{C}) \to (S \to \mathbb{C})$$

where the types $T = \mathbb{R}$ and $S = \mathbb{C}$. Note that the function type constructor $(\to)$ is used three times here: once in $T \to \mathbb{C}$, once in $S \to \mathbb{C}$ and finally at the top level to indicate that the transform maps functions to functions. This means that $\mathcal{L}$ is an example of a higher-order function, and we will see many uses of this idea in this book.

Now we move to introducing some of the ways types are defined in Haskell, our language of choice for the implementation (and often also specification) of mathematical concepts.

Figure 1.2: Humorously inappropriate type mismatch on a sign in New Cuyama, California. By I, MikeGogulski, CC BY 2.5, Wikipedia, 2021-12-27.

1.2   Types in Haskell: type, newtype, and data

There are three keywords in Haskell involved in naming and creating types: **type**, **newtype**, and **data**.

type – abbreviating type expressions   The **type** keyword is used to create a type synonym – just another name for a type expression. The new name is written on the left-hand side (LHS) of an equal sign, and the type expression on the right-hand side (RHS). The semantics is unchanged: the set of values of type *Number* is exactly the same as the set of values of type *Integer*, etc.

$$
\begin{aligned}
\textbf{type } & Number = Integer \\
\textbf{type } & Foo \quad\; = (Maybe\,[String],[[Number]]) \\
\textbf{type } & BinOp \;= Number \to Number \to Number \\
\textbf{type } & Env\,v\,s = [(v,s)]
\end{aligned}
$$

A **type** declaration does not add type safety, just readability (if used wisely). The *Env* example shows that a type synonym can have type parameters. Note that *Env v s* is a type (for any types *v* and *s*), but *Env* on its own is not a type but a *type constructor* — a function at the type level.

newtype – more protection   A simple example of the use of **newtype** in Haskell is to distinguish values which should be kept apart. A fun example of *not* keeping values apart is shown in Figure 1.2. To avoid this class of problems Haskell provides the **newtype** construct as a stronger version of **type**.

> **newtype** *Count*   = *Cou*   *Int*   -- Population count
> **newtype** *DistFeet* = *DisFt Int*   -- Elevation in feet above sea level
> **newtype** *Year*   = *Yea*   *Int*   -- Year of establishment
>   -- Example values of the new types
> *pop* :: *Count*;   *pop* = *Cou*   562;
> *hei* :: *DistFeet*; *hei* = *DisFt* 2150;
> *est* :: *Year*;   *est* = *Yea*   1951;

This example introduces three new types, *Count*, *DistFeet*, and *Year*, which all are internally represented by an *Int* but which are good to keep apart. The syntax also introduces *constructor functions Cou* :: *Int* $\to$ *Count*, *DisFt* and *Yea* which can be used to translate from plain integers to the new types, and for pattern matching. The semantics of *Count* is the set of values of the form *Cou i* for every value *i* :: *Int*. It is not the same as the semantics of *Int* but the sets are *bijective*. The function *Cou* is an invertible function, a *bijection*, also called a set-isomorphism. (We talk about isomorphisms between richer algebraic structures in Chapter 4.)

Later in this chapter we use a newtype for the semantics of complex numbers as a pair of numbers in the Cartesian representation but it may also be useful to have another newtype for complex as a pair of numbers in the polar representation.

The keyword data for syntax trees   The simplest form of a recursive datatype is the unary notation for natural numbers:

> **data** $N = Z \mid S\ N$

This declaration introduces

- a new type N for unary natural numbers,
- a constructor $Z :: N$ to represent zero, and
- a constructor $S :: N \to N$ to represent the successor function.

The semantics of N is the set of natural numbers ($\mathbb{N}$), with the semantics of Z being 0, $S\ Z$ being 1, etc. A way to be complete about the semantics is to state that the semantics of S is "add one".

Examples values: *zero* = Z, *one* = $S\ Z$, *three* = $S\ (S\ one)$.

The **data** keyword will be used throughout the book to define (inductive) datatypes of syntax trees for different kinds of expressions: simple arithmetic expressions, complex number expressions, etc. But it can also be used for non-inductive datatypes, like **data** $Bool = False \mid True$, or **data** $TwoDice = TD\ \mathbb{Z}\ \mathbb{Z}$. The $Bool$ type is the simplest example of a *sum type*, where each value uses either of the two variants $False$ and $True$ as the constructor. The $TwoDice$ type is an example of a *product type*, where each value uses the same constructor TD and records values for the values of two rolled dice. (See Exercise 1.3 for the intuition behind the terms "sum" and "product" used here.)

Maybe **and parameterised types**   It is very often possible to describe a family of types using a type parameter. One simple example is the type constructor *Maybe*, used for encapsulation of an optional value:

> **data** $Maybe\ a = Nothing \mid Just\ a$

This declaration introduces

- a new type $Maybe\ a$ for every type a,

- a constructor $Nothing :: Maybe\ a$ to represent "no value", and

- a constructor $Just :: a \rightarrow Maybe\ a$ to represent "just a value".

A maybe type can be used when an operation may, or may not, return a value:

> $inv :: \mathbb{Q} \rightarrow Maybe\ \mathbb{Q}$
> $inv\ 0 = Nothing$
> $inv\ r = Just\ (1\ /\ r)$

Two other examples of, often used, parameterised types are (a, b) for the type of pairs (a product type) and $Either\ a\ b$ for either an a or a b (a sum type). For reference, the either type is defined as follows in Haskell:

> **data** $Either\ p\ q = Left\ p \mid Right\ q$

1.3   Notation and abstract syntax for sequences

As preparation for the language of sequences and limits later (Sections 2.5.2 and 5.4), we spend a few lines on the notation and abstract syntax of sequences.

In maths textbooks, the following notation is commonly in use: $\{a_i\}_{i=0}^{\infty}$ or just $\{a_i\}$ and (not always) an indication of the type X of the a_i. Note that the a

at the centre of this notation actually carries all of the information: an infinite family of values a_i each of type X. If we interpret the subscript notation a_i as function application ($a(i)$) we can see that $a : \mathbb{N} \to X$ is a useful typing of an infinite sequence. Some examples:

type $\mathbb{N}$ $= Natural$ -- imported from *Numeric.Natural*
type $\mathbb{Q}^+$ $= Ratio\ \mathbb{N}$ -- imported from *Data.Ratio*
type $Seq\ a = \mathbb{N} \to a$

$idSeq :: Seq\ \mathbb{N}$
$idSeq\ i = i$ -- $\{0, 1, 2, 3, \ldots\}$
$invSeq :: Seq\ \mathbb{Q}^+$
$invSeq\ i = 1\ \%\ (1 + i)$ -- $\{\frac{1}{1}, \frac{1}{2}, \frac{1}{3}, \frac{1}{4}, \ldots\}$
$pow2 :: Num\ r \Rightarrow Seq\ r$
$pow2 = (2\hat{\ })$ -- $\{1, 2, 4, 8, \ldots\}$
$conSeq :: a \to Seq\ a$
$conSeq\ c\ i = c$ -- $\{c, c, c, c, \ldots\}$

What operations can be performed on sequences? We have seen the first one: given a value c we can generate a constant sequence with $conSeq\ c$. We can also add sequences componentwise (also called "pointwise"):

$addSeq :: Num\ a \Rightarrow Seq\ a \to Seq\ a \to Seq\ a$
$addSeq\ f\ g\ i = f\ i + g\ i$

and in general we can lift any binary operation $op :: a \to b \to c$ to the corresponding, pointwise, operation of sequences:

$liftSeq_2 :: (a \to b \to c) \to Seq\ a \to Seq\ b \to Seq\ c$
$liftSeq_2\ op\ f\ g\ i = op\ (f\ i)\ (g\ i)$ -- $\{op\ (f\ 0)\ (g\ 0), op\ (f\ 1)\ (g\ 1), \ldots\}$

Similarly we can lift unary operations, and "nullary" operations:

$liftSeq_1 :: (a \to b) \to Seq\ a \to Seq\ b$
$liftSeq_1\ h\ f\ i = h\ (f\ i)$ -- $\{h\ (f\ 0), h\ (f\ 1), h\ (f\ 2), \ldots\}$
$liftSeq_0 :: a \to Seq\ a$
$liftSeq_0\ c\ i = c$

Exercise 1.8: what does function composition do to a sequence? For a sequence a what is $a \circ (1+)$? What is $(1+) \circ a$?

Another common mathematical operator on sequences is the limit (of a sequence). We will get back to limits later (in Section 2.5.2), but for now we just note that an arbitrary sequence x, may or may not have a limit. One way to capture that idea is to let the return type be *Maybe X*, with *Nothing* corresponding to divergence. One way of typing a limit operator is $lim : (\mathbb{N} \to X) \to Maybe\ X$, but note that with this type we cannot implement the operator, because that would take infinitely long time.

Here we just specify one more common operation: the sum of a sequence (like $\sigma = \sum_{i=0}^{\infty} 1/i!$[1]). Just as not all sequences have a limit, not all have a sum either. But for every sequence we can define a new sequence of partial sums:

$$sums :: Num\ a \Rightarrow Seq\ a \to Seq\ a$$
$$sums\ a\ 0 = 0$$
$$sums\ a\ i = sums\ a\ (i-1) + a\ i$$

The function *sums* is perhaps best illustrated by examples:

$$sums\ (conSeq\ c) \ \texttt{==}\ \{0, c, 2*c, 3*c, \ldots\}$$
$$sums\ (idSeq) \qquad \texttt{==}\ \{0, 0, 1, 3, 6, 10, \ldots\}$$

The general pattern is to start at zero and accumulate the sum of initial prefixes of the input sequence.

By combining *sums* with limits we can state formally that the sum of an infinite sequence *a* exists and is *S* iff the limit of *sums a* exists and is *S*. We can write the above as a formula: *Just S = lim (sums a)*. For our example it turns out that the sum converges and that $\sigma = \sum_{i=0}^{\infty} 1/i! = e$ but we will not get to that until Section 8.1.

We will also return to (another type of) limits in Section 3.2 about derivatives where we explore variants of the classical definition

$$f'(x) = \lim_{h \to 0} \frac{f(x+h) - f(x)}{h}$$

To sum up this subsection, we have defined a small Domain-Specific Language (DSL) for infinite sequences by defining a type (*Seq a*), some operations (*conSeq, addSeq, liftSeq$_1$, sums, ...*) and some evaluation functions or predicates (like *lim* and *sum*).

1.4   A DSL of complex numbers

This section is partly based on material by Ionescu and Jansson [2016]. and we collect our definitions in a Haskell module which is available in the GitHub repository of the book.

---

[1] Here $n! = 1 * 2 * \ldots * n$ is the factorial .

module *DSLsofMath.ComplexSem* **where**

These definitions together form a DSL for complex numbers.

We now turn to our first study of mathematics as found "in the wild": we will do an analytic reading of a piece of the introduction of complex numbers by Adams and Essex [2010]. We choose a simple domain to allow the reader to concentrate on the essential elements of our approach without the distraction of potentially unfamiliar mathematical concepts. In fact, for this section, we temporarily pretend to forget any previous knowledge of complex numbers, and study the textbook as we would approach a completely new domain, even if that leads to a somewhat exaggerated attention to detail.

Adams and Essex introduce complex numbers in Appendix A of their book. The section *Definition of Complex Numbers* starts with:

> We begin by defining the symbol i, called **the imaginary unit**, to have the property
>
> $$i^2 = -1$$
>
> Thus, we could also call i the square root of -1 and denote it $\sqrt{-1}$. Of course, i is not a real number; no real number has a negative square.

At this stage, it is not clear what the type of i is meant to be, we only know that i is not a real number. Moreover, we do not know what operations are possible on i, only that i^2 is another name for -1 (but it is not obvious that, say $i * i$ is related in way with i^2, since the operations of multiplication and squaring have only been introduced so far for numerical types such as $\mathbb{N}$ or $\mathbb{R}$, and not for "symbols").

For the moment, we introduce a type for the symbol i, and, since we know nothing about other symbols, we make i the only member of this type. We use a capital I in the **data** declaration because a lowercase constructor name is a syntax error in Haskell, but for convenience we add also a value $i = I$.

data *ImagUnits* $= I$
i :: *ImagUnits*
i $= I$

We can give the translation from the abstract syntax to the concrete syntax as a function *showIU*:

showIU :: *ImagUnits* $\rightarrow$ *String*
showIU *I* $=$ `"i"`

Next, in the book, we find the following definition:

> **Definition: A complex number** is an expression of the form
>
> $$a + bi \qquad \text{or} \qquad a + ib,$$
>
> where a and b are real numbers, and i is the imaginary unit.

This definition clearly points to the introduction of a syntax (notice the keyword "form"). This is underlined by the presentation of *two* forms, which can suggest that the operation of juxtaposing i (multiplication?) is not commutative.

A profitable way of dealing with such concrete syntax in functional programming is to introduce an abstract representation of it in the form of a datatype:

$$\textbf{data } ComplexA = CPlus_1 \; \mathbb{R} \; \mathbb{R} \; ImagUnits \quad \text{-- the form } a + bi$$
$$\qquad\qquad\qquad | \;\; CPlus_2 \; \mathbb{R} \; ImagUnits \; \mathbb{R} \quad \text{-- the form } a + ib$$

We can give the translation from the (abstract) syntax to its concrete representation as a string of characters, as the function $showCA$:

$$showCA :: ComplexA \rightarrow \quad String$$
$$showCA \quad (CPlus_1 \; x \; y \; i) = show \; x \mathbin{+\!\!+} \text{" + "} \mathbin{+\!\!+} show \; y \mathbin{+\!\!+} showIU \; i$$
$$showCA \quad (CPlus_2 \; x \; i \; y) = show \; x \mathbin{+\!\!+} \text{" + "} \mathbin{+\!\!+} showIU \; i \mathbin{+\!\!+} show \; y$$

Notice that the type $\mathbb{R}$ is not implemented yet (and it is not even clear how to implement it with fidelity to mathematical convention at this stage) but we want to focus on complex numbers so we will simply approximate $\mathbb{R}$ by double precision floating point numbers for now.

$$\textbf{type } \mathbb{R} = Double$$

Adams and Essex continue with examples:

> For example, $3 + 2i$, $\frac{7}{2} - \frac{2}{3}i$, $i\pi = 0 + i\pi$ and $-3 = -3 + 0i$ are all complex numbers. The last of these examples shows that every real number can be regarded as a complex number.

The second example is somewhat problematic: it does not seem to be of the form $a + bi$. Given that the last two examples seem to introduce shorthand for various complex numbers, let us assume that this one does as well, and that $a - bi$ can be understood as an abbreviation of $a + (-b) \, i$. With this provision, in our Haskell encoding the examples are written as in Table 1.1. We

Mathematics		Haskell
$3 + 2i$		$CPlus_1$ 3 2 I
$\frac{7}{2} - \frac{2}{3}i$	$= \frac{7}{2} + \frac{-2}{3}i$	$CPlus_1$ (7 / 2) (−2 / 3) I
$i\pi$	$= 0 + i\pi$	$CPlus_2$ 0 I π
-3	$= -3 + 0i$	$CPlus_1$ (−3) 0 I

Table 1.1: Examples of notation and abstract syntax for some complex numbers.

interpret the sentence "The last of these examples ..." to mean that there is an embedding of the real numbers in *ComplexA*, which we introduce explicitly:

$toComplex :: \mathbb{R} \rightarrow ComplexA$
$toComplex\ x = CPlus_1\ x\ 0\ I$

Again, at this stage there are many open questions. For example, we can assume that the mathematical expression $i1$ stands for the complex number $CPlus_2$ 0 I 1, but what about the expression i by itself? If juxtaposition is meant to denote some sort of multiplication, then perhaps 1 can be considered as a unit, in which case we would have that i abbreviates $i1$ and therefore $CPlus_2$ 0 I 1. But what about, say, $2i$? Abbreviations with i have only been introduced for the ib form, and not for the bi one!

The text then continues with a parenthetical remark which helps us dispel these doubts:

> (We will normally use $a + bi$ unless b is a complicated expression, in which case we will write $a + ib$ instead. Either form is acceptable.)

This remark suggests strongly that the two syntactic forms are meant to denote the same elements, since otherwise it would be strange to say "either form is acceptable". After all, they are acceptable according to the definition provided earlier.

Given that $a + ib$ is only "syntactic sugar" for $a + bi$, we can simplify our representation for the abstract syntax, merging the two constructors:

data *ComplexB* $= CPlusB\ \mathbb{R}\ \mathbb{R}\ ImagUnits$

In fact, since it doesn't look as though the type *ImagUnits* will receive more elements, we can dispense with it altogether:

data *ComplexC* $= CPlusC\ \mathbb{R}\ \mathbb{R}$

(The renaming of the constructor to *CPlusC* serves as a guard against the case that we have suppressed potentially semantically relevant syntax.)

We read further:

> It is often convenient to represent a complex number by a single letter; w and z are frequently used for this purpose. If a, b, x, and y are real numbers, and $w = a + bi$ and $z = x + yi$, then we can refer to the complex numbers w and z. Note that $w = z$ if and only if $a = x$ and $b = y$.

First, let us notice that we are given an important semantic information: to check equality for complex numbers, it is enough to check equality of the components (the arguments to the constructor *CPlusC*). Another way of saying this is that *CPlusC* is injective. In Haskell we could define this equality as:

> **instance** *Eq ComplexC* **where**
>   *CPlusC a b* == *CPlusC x y* = a == $x \wedge b$ == y

The line **instance** *Eq ComplexC* is there to explain to Haskell that *ComplexC* supports the (==) operator. (The cognoscenti would prefer to obtain an equivalent definition using the shorter **deriving** *Eq* clause upon defining the type.)

This shows that the set of complex numbers is, in fact, isomorphic with the set of pairs of real numbers, a point which we can make explicit by reformulating the definition in terms of a **newtype**:

> **newtype** *ComplexD* = *CD* $(\mathbb{R}, \mathbb{R})$ **deriving** *Eq*

As we see it, the somewhat confusing discussion of using "letters" to stand for complex numbers serves several purposes. First, it hints at the implicit typing rule that the symbols z and w should be complex numbers. Second, it shows that, in mathematical arguments, one needs not abstract over two real variables: one can instead abstract over a single complex variable. We already know that we have an isomorphism between pair of reals and complex numbers. But additionally, we have a notion of *pattern matching*, as in the following definition:

> **Definition:** If $z = x + yi$ is a complex number (where x and y are real), we call x the **real part** of z and denote it $Re\ (z)$. We call y the **imaginary part** of z and denote it $Im\ (z)$:
>
> $$Re\ (z) = Re\ (x + yi) = x$$
> $$Im\ (z) = Im\ (x + yi) = y$$

This is rather similar to Haskell's *as-patterns*:

$$re :: ComplexD \quad \rightarrow \mathbb{R}$$
$$re \; z@(CD \; (x,y)) \; = x$$
$$im :: ComplexD \quad \rightarrow \mathbb{R}$$
$$im \; z@(CD \; (x,y)) \; = y$$

a potential source of confusion being that the symbol z introduced by the as-pattern is not actually used on the right-hand side of the equations (although it could be).

The use of as-patterns such as "$z = x + yi$" is repeated throughout the text, for example in the definition of the algebraic operations on complex numbers:

The sum and difference of complex numbers

If $w = a + bi$ and $z = x + yi$, where a, b, x, and y are real numbers, then

$$w + z = (a + x) + (b + y) \; i$$
$$w - z = (a - x) + (b - y) \; i$$

With the introduction of algebraic operations, the domain-specific language of complex numbers becomes much richer. We can describe these operations in a *shallow embedding* in terms of the concrete datatype *ComplexD*, for example:

$$addD :: ComplexD \rightarrow ComplexD \rightarrow ComplexD$$
$$addD \; (CD \; (a,b)) \; (CD \; (x,y)) = CD \; ((a+x),(b+y))$$

or we can build a datatype of "syntactic" complex numbers from the algebraic operations to arrive at a *deep embedding* as seen in the next section. Both shallow and deep embeddings will be further explained in Sections 1.7.1 and 4.6 (and in several other places: this is a recurrent idea of the book).

At this point we can sum up the "evolution" of the datatypes introduced so far. Starting from *ComplexA*, the type has evolved by successive refinements through *ComplexB*, *ComplexC*, ending up in *ComplexD* (see Fig. 1.3). We can also make a parameterised version of *ComplexD*, by noting that the definitions for complex number operations work fine for a range of underlying numeric types. The operations for *ComplexSem* are defined in module *CSem*, available in Appendix B.

$$
\begin{array}{lll}
\textbf{data} & \textit{ImagUnits} & = I \\
\textbf{data} & \textit{ComplexA} & = \textit{CPlus}_1 \; \mathbb{R} \; \mathbb{R} \; \textit{ImagUnits} \\
& & \mid \; \textit{CPlus}_2 \; \mathbb{R} \; \textit{ImagUnits} \; \mathbb{R} \\
\textbf{data} & \textit{ComplexB} & = \textit{CPlusB} \; \mathbb{R} \; \mathbb{R} \; \textit{ImagUnits} \\
\textbf{data} & \textit{ComplexC} & = \textit{CPlusC} \; \mathbb{R} \; \mathbb{R} \\
\textbf{newtype} \; \textit{ComplexD} & & = \textit{CD} \; (\mathbb{R}, \mathbb{R}) \; \textbf{deriving} \; \textit{Eq} \\
\textbf{newtype} \; \textit{ComplexSem} \; r & = \textit{CS} \; (r, r) & \textbf{deriving} \; \textit{Eq}
\end{array}
$$

Figure 1.3: Complex number datatype refinement (semantics).

1.5 A syntax for (complex) arithmetical expressions

By following Adams and Essex [2010], we have arrived at a representation which captures the *semantics* of complex numbers. This kind of representation is often called a "shallow embedding". Now we turn to the study of the *syntax* instead ("deep embedding"). We collect these syntactic definitions in a separate module which imports the earlier semantic definitions.

> **module** *DSLsofMath.ComplexSyn* **where**
> **import** *DSLsofMath.CSem* (*Complex* (*C*), *addC*, *mulC*, *Ring*)
> **import** *DSLsofMath.ComplexSem*

We want a datatype *ComplexE* for the abstract syntax tree (AST) of expressions (a DSL for complex arithmetical expressions). The syntactic expressions can later be evaluated to semantic values. The concept of "an evaluator", a function from the syntax to the semantics, is something we will return to many times in this book.

> $evalE :: ComplexE \rightarrow ComplexD$

The datatype *ComplexE* should collect ways of building syntactic expressions representing complex numbers and we have so far seen the symbol *i*, an embedding from $\mathbb{R}$, addition and multiplication. We make these four operations into *constructors* in one recursive datatype as follows:

> $\textbf{data} \; ComplexE = I$
> $\qquad\qquad\quad \mid \; ToComplex \; \mathbb{R}$
> $\qquad\qquad\quad \mid \; Add \; ComplexE \; ComplexE$
> $\qquad\qquad\quad \mid \; Mul \; ComplexE \; ComplexE$
> $\textbf{deriving} \; (Eq, Show)$

Note that, in *ComplexA* above, we also had a constructor for addition (*CPlus*$_1$), but it was playing a different role. They are distinguished by type: *CPlus*$_1$

took two real numbers as arguments, while *Add* here takes two complex expressions as arguments.

Here are two examples of type *ComplexE* as Haskell code and as ASTs:

$$testE1 = Mul\ I\ I$$
$$testE2 = Add\ (ToComplex\ 3)\ (Mul\ (ToComplex\ 2)\ I)$$

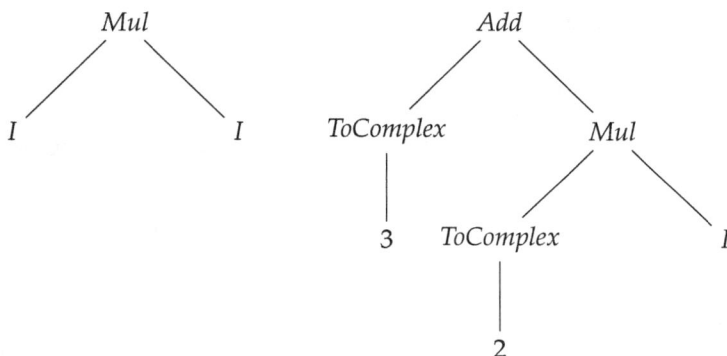

We can implement the evaluator *evalE* by pattern matching on the constructors of the syntax tree and by recursion. To write a recursive function requires a small leap of faith. It can be difficult to get started implementing a function (like *evalE*) that should handle all the cases and all the levels of a recursive datatype (like *ComplexE*). One way to overcome this difficulty is through what may seem at first glance "wishful thinking": assume that all but one case have been implemented already. All you need to do is to focus on that one remaining case, and you can freely call the function (that you are implementing) recursively, as long as you do it for subexpressions (subtrees of the abstract syntax tree datatype). This pattern is called *structural induction*.

For example, when implementing the *evalE* (*Add* c_1 c_2) case, you can assume that you already know the values $s_1, s_2 :: ComplexD$ corresponding to the subtrees c_1 and c_2 of type *ComplexE*. The only thing left is to add them up componentwise and we can assume there is a function *addD* :: *ComplexD* → *ComplexD* → *ComplexD* taking care of this step (in fact, we implemented it earlier in Section 1.4). Continuing in this direction (by structural induction; or "wishful thinking") we arrive at the following implementation.

```
evalE I              = iD
evalE (ToComplex r)  = toComplexD r
evalE (Add c₁ c₂)    = addD (evalE c₁) (evalE c₂)
evalE (Mul c₁ c₂)    = mulD (evalE c₁) (evalE c₂)
```

Note the pattern here: for each constructor of the syntax datatype we assume that there exists a corresponding semantic function. The next step is to implement these functions, but let us first list their types and compare them with the types of the syntactic constructors:

$$I \ :: ComplexE$$
$$iD :: ComplexD$$

$$ToComplex \ :: \mathbb{R} \to ComplexE$$
$$toComplexD :: \mathbb{R} \to ComplexD$$

$$Mul \ :: ComplexE \ \to ComplexE \ \to ComplexE$$
$$mulD :: ComplexD \to ComplexD \to ComplexD$$

As we can see, each use of *ComplexE* has been replaced be a use of *ComplexD*. Finally, we can start filling in the implementations:

$$iD \qquad = CD \ (0,1)$$
$$toComplexD \ r = CD \ (r,0)$$

The function *addD* was defined earlier and *mulD* is left as an exercise for the reader. To sum up we have now implemented a recursive datatype for mathematical expressions describing complex numbers, and an evaluator that computes the underlying number. Note that many different syntactic expressions will evaluate to the same number (*evalE* is not injective).

Generalising from the example of *testE2* we also define a function to embed a semantic complex number in the syntax:

$$fromCD :: ComplexD \to ComplexE$$
$$fromCD \ (CD \ (x,y)) = Add \ (ToComplex \ x) \ (Mul \ (ToComplex \ y) \ I)$$

This function is injective: different complex numbers map to different syntactic expressions.

1.6   Laws, properties and testing

There are certain laws that we would like to hold for operations on complex numbers. To specify these laws, in a way which can be easily testable in Haskell, we use functions to *Bool* (also called *predicates* or *properties*). The intended meaning of such a Boolean function (representing a law) is "for all inputs, this should return *True*". This idea is at the core of *property based testing* (pioneered by Claessen and Hughes [2000]) and conveniently available in the Haskell library QuickCheck.

Note that a predicate $p : A \to Bool$ can also be used to specify the subset of A for which p returns *True*. QuickCheck is very good at finding counterexamples to candidate laws: values for which p returns *False*. With the subset interpretation this means that QuickCheck helps finding elements in the set specified by the opposite predicate $\neg \circ p : A \to Bool$. When it fails, we hope that the set of counterexamples is actually empty, but we cannot know that for sure without a proof (or by exhaustive testing of all cases).

The simplest law is perhaps $i^2 = -1$ from the start of Section 1.4,

> *propI2* :: *Bool*
> *propI2* = *Mul I I* === *ToComplex* (-1)

Note the we use a new operator here, (===). Indeed, we reserve the usual equality (==) for syntactic equality (and here the left hand side (LHS) is clearly not syntactically equal to the right hand side). The new operator (===) corresponds to semantic equality, that is, equality *after evaluation*:

> (===) :: *ComplexE* $\to$ *ComplexE* $\to$ *Bool*
> z === w $\quad = \quad$ *evalE* z == *evalE* w

Another law is that *fromCD* is an embedding: if we start from a semantic value, embed it back into syntax, and evaluate that syntax we get back to the value we started from.

> *propFromCD* :: *ComplexD* $\to$ *Bool*
> *propFromCD* s = *evalE* (*fromCD* s) == s

Other desirable laws are that $(+)$ and $(*)$ should be associative and commutative and $(*)$ should distribute over $(+)$:

> | | | | | |
> |---|---|---|---|---|
> | *propCommAdd* | $x\ y$ | $=$ | $x + y$ | === $y + x$ |
> | *propCommMul* | $x\ y$ | $=$ | $x * y$ | === $y * x$ |
> | *propAssocAdd* | $x\ y\ z$ | $=$ | $(x + y) + z$ | === $x + (y + z)$ |
> | *propAssocMul* | $x\ y\ z$ | $=$ | $(x * y) * z$ | === $x * (y * z)$ |
> | *propDistMulAdd* $x\ y\ z$ | | $=$ | $x * (y + z)$ | === $(x * y) + (x * z)$ |

These laws actually fail, but not due to any mistake in the implementation of *evalE* in itself. To see this, let us consider associativity at different types:

> *propAssocInt* $\quad$ = *propAssocAdd* :: *Int* $\quad \to$ *Int* $\quad \to$ *Int* $\quad \to$ *Bool*
> *propAssocDouble* = *propAssocAdd* :: *Double* $\to$ *Double* $\to$ *Double* $\to$ *Bool*

The first property is fine, but the second fails. Why? QuickCheck can be used to find small examples — this one is perhaps the best one:

$$notAssocEvidence :: (Double, Double, Double, Bool)$$
$$notAssocEvidence = (lhs, rhs, lhs - rhs, lhs \mathrel{==} rhs)$$
$$\textbf{where } lhs = (1 + 1) + 1 / 3$$
$$rhs = 1 + (1 + 1 / 3)$$

For completeness: these are the values:

(2.3333333333333335	-- Notice the five at the end
, 2.333333333333333,	-- which is not present here.
, 4.440892098500626e−16	-- The (very small) difference
, *False*)	

We can now see the underlying reason why some of the laws failed for complex numbers: the approximative nature of *Double*. Therefore, to ascertain that there is no other bug hiding, we need to move away from the implementation of $\mathbb{R}$ as *Double*. We do this by abstraction: we make one more version of the complex number type, which is parameterised on the underlying representation type for $\mathbb{R}$. At the same time, to reduce the number of constructors, we combine *I* and *ToComplex* to *ToComplexCart*, which corresponds to the primitive form $a + bi$ discussed above:

$$\textbf{data } ComplexSyn\ r = ToComplexCart\ r\ r$$
$$\mid\ ComplexSyn\ r \mathrel{:+:} ComplexSyn\ r$$
$$\mid\ ComplexSyn\ r \mathrel{:*:} ComplexSyn\ r$$

$$toComplexSyn :: Num\ a \Rightarrow a \rightarrow ComplexSyn\ a$$
$$toComplexSyn\ x = ToComplexCart\ x\ 0$$

From Appendix B we import **newtype** *Complex r* = *C* (r, r) **deriving** *Eq* and the semantic operations *addC* and *mulC* corresponding to *addD* and *mulD*.

$$evalCSyn :: Ring\ r \Rightarrow ComplexSyn\ r \rightarrow Complex\ r$$
$$evalCSyn\ (ToComplexCart\ x\ y) = C\ (x, y)$$
$$evalCSyn\ (l \mathrel{:+:} r) \qquad = addC\ (evalCSyn\ l)\ (evalCSyn\ r)$$
$$evalCSyn\ (l \mathrel{:*:} r) \qquad = mulC\ (evalCSyn\ l)\ (evalCSyn\ r)$$

With this parameterised type we can test the code for "complex rationals" to avoid rounding errors. (The reason why maths textbooks rarely talk about complex rationals is because complex numbers are used to handle roots of all numbers uniformly, and roots are in general irrational.)

1.6.1   Generalising laws

Some laws appear over and over again in different mathematical contexts. For example, binary operators are often associative or commutative, and sometimes one operator distributes over another. We will work more formally with logic in Chapter 2 but we introduce a few definitions already here:

Associative $(\circledast) = \forall\, a, b, c.\ (a \circledast b) \circledast c = a \circledast (b \circledast c)$

Commutative $(\circledast) = \forall\, a, b.\ a \circledast b = b \circledast a$

Distributive $(\otimes)\,(\oplus) = \forall\, a, b, c.\ (a \oplus b) \otimes c = (a \otimes c) \oplus (b \otimes c)$

The above laws are *parameterised* over some operators $((\circledast), (\otimes), (\oplus))$. These laws will hold for some operators, but not for others. For example, division is not commutative; taking the average of two quantities is commutative but not associative. (See also Item 4, on page 106 for further analysis of distributivity.) Such generalisations can be reflected in QuickCheck properties as well.

$$propAssoc :: SemEq\ a \Rightarrow (a \to a \to a) \to a \to a \to a \to Bool$$
$$propAssoc\ (\circledast)\ x\ y\ z = (x \circledast y) \circledast z \mathbin{=\!=\!=} x \circledast (y \circledast z)$$

Note that *propAssocA* is a higher-order function: it takes a function $(\circledast)$ (declared as a binary operator) as its first parameter, and tests if it is associative. The property is also polymorphic: it works for many different types a (all types which have an $=\!=\!=$ operator).

Thus we can specialise it to *Add*, *Mul* and any other binary operator, and obtain some of the earlier laws (*propAssocAdd*, *propAssocMul*). The same can be done with distributivity. Doing so we learnt that the underlying set matters: $(+)$ for $\mathbb{R}$ has some properties, but $(+)$ for *Double* has others. When formalising maths as DSLs, approximation is sometimes convenient, but makes many laws false. Thus, we should attempt to do it late, and if possible, leave a parameter to make the degree of approximation tunable (*Int*, *Integer*, *Float*, *Double*, $\mathbb{Q}$, syntax trees, etc.). For the curious, some of our more advanced material about testing and different kinds of polymorphism is available in [Bernardy et al., 2010, Ionescu and Jansson, 2011, Jeuring et al., 2012].

1.7 Types of functions, expressions and operators

We start this section with a common pitfall with traditional mathematical notation. Mathematical texts often talk about "the function $f(x)$" when "the function f" would be more clear. Otherwise there is a risk of confusion between $f(x)$ as a function and $f(x)$ as the value you get from applying the function f to the value bound to the name x.

Examples: let $f(x) = x - 1$ and let $t = 5 * f(2)$. Then it is clear that the value of t is the constant 15. But if we let $s = 5 * f(x)$ it is not clear if s should be seen as a constant (for some fixed value x) or as a function of x.

Paying attention to types and variable scope often helps to sort out these ambiguities and we will expand on this in the rest of this chapter.

Examples of types in mathematics   Simple types are sometimes mentioned explicitly in mathematical texts:

- $x \in \mathbb{R}$

- $\sqrt{} : \mathbb{R}_{\geq 0} \to \mathbb{R}_{\geq 0}$

- $(\_)^2 : \mathbb{R} \to \mathbb{R}$ or, alternatively but *not* equivalently

- $(\_)^2 : \mathbb{R} \to \mathbb{R}_{\geq 0}$

However the types of big operators (sums, limits, integrals, etc.) are usually not given explicitly. In fact, it may not be clear at first sight that the summing operator ($\sum$) should be assigned a type at all! Yet this is exactly what we will set out to do, dealing with the pitfalls of mathematical notation just introduced. However, to be able to do so convincingly we shall first clarify the relationship between functions and expressions.

1.7.1   Expressions and functions of one variable

Consider the following examples of mathematical function definitions:

- $f(x) = x - 1$

- $g(x) = 2 * x^2 + 3$

- $h(y) = 2 * y^2 + 3$

As the reader may guess by now, we can assign to f, g, h the type $\mathbb{R} \to \mathbb{R}$. Other choices could work, such as $\mathbb{Z} \to \mathbb{Z}$, etc., but for sure, they are functions. Additionally, the name of the variable appears to play no role in the meaning of the functions, and we can say, for example, $g = h$.

Consider now the three expressions:

- $e_1 = x - 1$

- $e_2 = 2 * x^2 + 3$

- $e_3 = 2 * y^2 + 3$

These are all expressions of one (free) variable. We could say that their type is $\mathbb{R}$ — assuming that the free variable also has type $\mathbb{R}$. Furthermore, it is less clear now if $e_2 = 2 * x + 3 \stackrel{?}{=} 2 * y + 3 = e_3$. In general one cannot simply change a variable name to another without making sure that: 1. the renaming is applied everywhere uniformly and 2. the new variable name is not used for another purpose in the same scope (otherwise one informally says that there is a "name clash").

To clarify this situation, we will now formalise expressions of one variables as a DSL. For simplicity we will focus on arithmetic expressions only. Therefore we have constructors for addition, multiplication and constants, as in Section 1.5. Additionally, we have the all-important constructor for variables, which we will call X here. We can implement all this in a datatype as follows:

Deep embedding

```
data FunExp = Const ℝ
            | X
            | FunExp :+: FunExp
            | FunExp :*: FunExp
```

We could encode our examples as follows:

- $e_1 = X :+: Const\ (-1)$

- $e_2 = Const\ 2 :*: (X :*: X) :+: Const\ 3$

We no longer have a third example: in this type we can only ever represent one variable, as X, and thus we skip the last example, equal to the second.

We can now evaluate the values of these expressions. The meaning of operators and constants is as in Section 1.5. But, to be able to evaluate X, the variable, we need its value — and we simply take it as a parameter.

```
eval :: FunExp → ℝ → ℝ
eval  (Const α)  x  = α
eval  X          x  = x
eval  (e₁ :+: e₂) x  = eval e₁ x + eval e₂ x
eval  (e₁ :*: e₂) x  = eval e₁ x * eval e₂ x
```

However, we can make an equivalent interpretation of the above type as

$$eval :: FunExp \to (\mathbb{R} \to \mathbb{R})$$

That is, *FunExp* can be interpreted as a function! This is perhaps surprising, but the reason is that we used a fixed Haskell symbol (constructor) for the variable. There is only a single variable available in the syntax of *FunExp*, and thus such expressions are really equivalent to functions of a single argument.

Shallow embedding

Thus the above was a deep embedding for functions of a single variable. A shallow embedding would be using functions as the representation, say:

> **type** $FunExpS = \mathbb{R} \to \mathbb{R}$

Then we can define the operators directly on functions, as follows:

> $funConst\ \alpha = \lambda x \to \alpha$ -- thus $funConst = const$
> $funX\qquad = \lambda x \to x$ -- and $funX = id$
> $funAdd\ f\ g = \lambda x \to f\ x + g\ x$
> $funMul\ f\ g = \lambda x \to f\ x * g\ x$

Again, we have two possible intuitive readings of the above equations. The first reading, as expressions of a single variable, is that a constant α is interpreted as a constant function; the variable (X) is interpreted as the identity function; the sum of two expressions is interpreted as the sum of the evaluation of the operands, etc.

The second reading is that we can define an arithmetic structure $(*, +,$ etc.) on functions, by lifting the operators to work pointwise (as we did in Section 1.3).

To wrap it up, if we are so inclined, we can redefine the evaluator of the deep embedding using the operators of the shallow embedding:

> $eval :: FunExp\quad \to (\mathbb{R} \to \mathbb{R})$
> $eval\quad (Const\ \alpha)\ = funConst\ \alpha$
> $eval\quad X\qquad\qquad = funX$
> $eval\quad (e_1 :+: e_2) = funAdd\ (eval\ e_1)\ (eval\ e_2)$
> $eval\quad (e_1 :*: e_2) = funMul\ (eval\ e_1)\ (eval\ e_2)$

Representing expressions of one variable as functions (of one argument) is a recurring technique in this book. To start off, we can use it to assign types to big operators.

1.7.2 Scoping and typing big operators

We now introduce another common pitfall with traditional mathematical notation.

Scoping the integral operator   The syntax and scoping rules for the integral sign are rarely explicitly mentioned, but looking at it from a software perspective can help. If we start from a simple example, like $\int_1^2 x^2 dx$, it is relatively clear: the integral sign takes two real numbers as limits and then a certain notation for a function, or expression, to be integrated. Comparing the part after the integral sign to the syntax of a function definition $f(x) = x^2$ reveals a rather odd rule: instead of *starting* with declaring the variable x, the integral syntax *ends* with the variable name, and also uses the letter "d". (There are historical explanations for this notation, and it is motivated by computation rules in the differential calculus, but we will not go there now. We are also aware that the notation $\int dx f(x)$, which emphasises the bound variable, is sometimes used, especially by physicists, but it remains the exception rather than the rule at the time of writing.) It seems like the scope of the variable "bound" by d is from the integral sign to the final dx, but does it also extend to the limits of the domain of integration? The answer is no, as we can see from a slightly extended example:

$$f(x) = x^2$$
$$g(x) = \int_x^{2x} f(x)dx \qquad\qquad = \int_x^{2x} f(y)dy$$

The variable x bound on the left is independent of the variable x "bound under the integral sign". We address this issue in detail in the rest of this chapter. Mathematics textbooks usually avoid the risk of confusion by (silently) renaming variables when needed, but we believe that this renaming is a sufficiently important operation to be more explicitly mentioned.

The "big sum" operator   Consider the mathematical expression

$$\sum_{i=1}^n i^2$$

To be able to see which type is appropriate for $\sum$, we have to consider the type of the summand (i^2 in the example) first. As you may have guessed, it is an expression of one variable (i). You may object: but surely the body of the summation operator can use other variables! You would be entirely correct. However, *from the point of view of the summation*, it is as if such other variables were constant. Accepting this assertion as a fact until we can show a more complicated example, we can now assign a type to the summation operator. For simplicity, we will be using the shallow embedding; thus the operand can be typed as, say, $\mathbb{Z} \to R$. The other arguments will be the lower and upper limits of the sum (1 and n in our example). The variable name, i

shall not be represented as an argument: indeed, the variable name is *fixed by the representation of functions*. There is no choice to make at the point of summation. Thus, we write:

$$bigsum :: \mathbb{Z} \to \mathbb{Z} \to (\mathbb{Z} \to R) \to R$$

Conveniently, we can even provide a simple implementation:

$$bigsum\ low\ high\ f = sum\ [f\ i \mid i \leftarrow [low\mathinner{.\,.}high]]$$

and use it for our example as follows:

$$sumOfSquares\ n = bigsum\ 1\ n\ (\hat{}\ 2)$$

Equivalently, we can use a lambda expression for the summand, to give a name to the summation variable:

$$sumOfSquares\ n = bigsum\ 1\ n\ (\lambda i \to i \hat{}\ 2)$$

(Recall the syntax for lambda expressions from Section 1.1.2.)

As another example, let us represent the following nested sum

$$\sum_{i=1}^{m}\sum_{j=1}^{n} i^2 + j^2$$

using the shallow embedding of summation. This representation can be written simply as follows:

$$exampleSum\ m\ n = bigsum\ 1\ m\ (\lambda i \to bigsum\ 1\ n\ (\lambda j \to i \hat{}\ 2 + j \hat{}\ 2))$$

Aren't we cheating though? Surely we said that only one variable could occur in the summand, but we see both i and j? Well, we are not cheating as long as we use the *shallow embedding* for functions of one variable. Doing so allows us to: 1. use lambda notation to bind (and name) the variable name of the summation however we wish (in this case i and j) and 2. freely use any Haskell function of type $\mathbb{Z} \to R$ as the summand. In particular, this function can be any lambda expression returning $\mathbb{R}$, and this expression can include summation itself. This freedom is an advantage of shallow embeddings: if we were to use the deep embedding, then we would need much more machinery to ensure that we can represent summation within the deep embedding. In particular we need a way to embed variable binding itself. And we shall not be opening this can of worms just yet, even though we take a glimpse in Section 1.7.3.

Sticking conveniently to the shallow embedding, we can apply the same kind of reasoning to other big operators, and obtain the following typings:

- *lim* : $(\mathbb{N} \to \mathbb{R}) \to \mathbb{R}$ for the mathematical expression $\lim_{n\to\infty}\{a_n\}$.

- $\frac{d}{dt}$: $(\mathbb{R} \to \mathbb{R}) \to (\mathbb{R} \to \mathbb{R})$

  Note that there are many notations for derivatives. Instead of $\frac{d}{dt}f$ one sees also df/dt, or f' or even $\dot{f}$ if the variable is time (t). Below we will normally use the notation Df.

In sum, the chief difficulty to overcome when assigning types for mathematical operators is that they often introduce (bind) variable names. To take another example from the above, $\lim_{n\to\infty}$ binds n in a_n. In this book our stance is to make this binding clear by letting the body of the limit (a_n in the example) be a function. Thus we use the type $\mathbb{N} \to \mathbb{R}$ for the body. Therefore the limit operator is a higher-order function. A similar line of reasoning justifies the types of derivatives. We return to derivatives in Chapter 3.

1.7.3 Expressions of several variables

In a first reading this section can be skipped, however it is natural to extend the study of expressions from single variables to multiple variables.

The data type of expressions of multiple variables Let us define the following type, describing a deep embedding for simple arithmetic expressions. Compared to single variable expressions, we add one argument for variables, giving the *name* of the variable. Here we use a string, so we have an infinite supply of variables.

> **data** *MVExp* = *Va String* | *Ad MVExp MVExp* | *Di MVExp MVExp*

The last constructor *Di* is intended for division (for a change). Example expressions include $v = Va\ "v"$, $e_1 = Ad\ v\ v$, and $e_2 = Di\ e_1\ e_1$.

In the evaluator for $e :: MVExp$ we use an environment *env* to look up variables whose values are needed to compute the value of the expression. Because the semantics of *env* is a partial function (modelled as a total function of type *String* $\to$ *Maybe* $\mathbb{Q}$), the semantics of *MVExp* is a partial function too, of type *Env String* $\mathbb{Q} \to$ *Maybe* $\mathbb{Q}$.

> *evalMVExp* :: *MVExp* $\to$ *Env String* $\mathbb{Q} \to$ *Maybe* $\mathbb{Q}$
> *lookup*   :: *String* $\to$ *Env String* $\mathbb{Q} \to$ *Maybe* $\mathbb{Q}$

Note that there are two sources of *Nothing* in the evaluator: undefined variables, and (avoiding) division by zero.

To follow the "wishful thinking" pattern for the evaluator, we need helper functions for the three cases. We have presented *lookup* for variables earlier, and now we define *mayAdd* and *mayDiv* to handle the remaining constructors.

```
type Q = Rational
mayAdd :: Maybe Q → Maybe Q → Maybe Q
mayAdd (Just a) (Just b) = Just (a + b)
mayAdd _        _        = Nothing
mayDiv :: Maybe Q → Maybe Q → Maybe Q
mayDiv (Just a) (Just 0) = Nothing
mayDiv (Just a) (Just b) = Just (a / b)
mayDiv _        _        = Nothing
evalMVExp e env = eval e
  where
    eval (Va x)     = lookup x env    -- same as evalEnv env x
    eval (Ad e₁ e₂) = mayAdd (eval e₁) (eval e₂)
    eval (Di e₁ e₂) = mayDiv (eval e₁) (eval e₂)
```

The approach taken above is to use a *String* to name each variable: indeed, *Env String* Q is like a table of several variables and their values.

Polymorphic variables   In *MVExp* we used variables of type *String*, but it is often convenient to be able to choose another type for variables. Here is the same code for *eval* but with a type parameterised over the choice of type for the variables. Note that the type *PExp String* corresponds to *MVExp*.

```
data PExp v = V v | A (PExp v) (PExp v) | D (PExp v) (PExp v)
evalPExp :: Eq v ⇒ PExp v → Env v Q → Maybe Q
evalPExp e env = eval e
  where
    eval (V x)     = lookup x env
    eval (A e₁ e₂) = mayAdd (eval e₁) (eval e₂)
    eval (D e₁ e₂) = mayDiv (eval e₁) (eval e₂)
```

1.8 Exercises: Haskell, DSLs and expressions

Exercise 1.1. Consider the following data type for arithmetic expressions:

data $Exp = Con\ Integer$
$\qquad\quad |\ Plus\quad Exp\ Exp$
$\qquad\quad |\ Minus\ Exp\ Exp$
$\qquad\quad |\ Times\ \ Exp\ Exp$
$\qquad$ **deriving** $(Eq, Show)$

1. Write the following expressions in Haskell, using the Exp data type:

   (a) $a_1 = 2 + 2$

   (b) $a_2 = a_1 + 7 * 9$

   (c) $a_3 = 8 * (2 + 11) - (3 + 7) * (a_1 + a_2)$

2. Create a function $eval :: Exp \rightarrow Integer$ that takes a value of the Exp data type and returns the corresponding number (for instance, $eval\ a_2$ == 4). Try it on the expressions from the first part, and verify that it works as expected.

3. Consider the following mathematical expression with variables:

   $$c_1 = (x - 15) * (y + 12) * z$$
   $$\textbf{where } x = 5; y = 8; z = 13$$

   In order to represent this with our Exp data type, we are going to have to make some modifications:

   (a) Update the Exp data type with a new constructor $Var\ String$ that allows variables with strings as names to be represented. Use the updated Exp to write an expression for c_1 in Haskell.

   (b) Create a function $varVal :: String \rightarrow Integer$ that takes a variable name, and returns the value of that variable. For now, the function just needs to be defined for the variables in the expression above, i.e. $varVal$ "x" should return 5, $varVal$ "y" should return 8, and $varVal$ "z" should return 13.

   (c) Update the $eval$ function so that it supports the new Var constructor, and use it get a numeric value of the expression c_1.

Exercise 1.2. We will now look at a more general version of the *Exp* type from the previous exercise:

```
data E2 a = Con a
          | Var String
          | Plus  (E2 a) (E2 a)
          | Minus (E2 a) (E2 a)
          | Times (E2 a) (E2 a)
     deriving (Eq, Show)
```

The type has now been parameterised, so that it is no longer limited to representing expressions with integers, but can instead represent expressions with any type. For instance, we could have an *E2 Double* to represent expression trees with doubles at the leaves, or an *E2 ComplexD* to represent expression trees with complex numbers at the leaves.

1. Write the following expressions in Haskell, using the new *E2* data type.

   (a) $a_1 = 2.0 + a$

   (b) $a_2 = 5.3 + b * c$

   (c) $a_3 = a * (b + c) - (d + e) * (f + a)$

2. In order to evaluate these expressions, we will need a way of translating a variable name into the value. The following table shows the value of each variable in the expressions above:

Name	a	b	c	d	e	f
Value	1.5	4.8	2.4	7.4	5.8	1.7

   In Haskell, we can represent this table using a value of type *Table a = Env String a = [(String, a)]*, which is a list of pairs of variable names and values, where each entry in the list corresponds to a column in the table.

   (a) Express the table above in Haskell by creating *vars :: Table Double*.

   (b) Create a function *varVal :: Table a → String → a* that returns the value of a variable, given a table and a variable name. For instance, *varVal vars "d"* should return 7.4

   (c) Create a function *eval :: Num a ⇒ Table a → E2 a → a* that takes a value of the new *E2* data type and returns the corresponding number. For instance, *eval vars (Plus (Con 2) (Var "a"))* == 3.5. Try it on the expressions from the first part, and verify that it works as expected.

Exercise 1.3. [**Important**] Counting values. Assume we have three finite types *a, b, c* with cardinalities *A, B,* and *C.* (For example, the cardinality of *Bool* is 2, the cardinality of *Weekday* is 7, etc.) Then what is the cardinality of the types *Either a b*? *(a, b)*? *a → b*? etc. These rules for computing the cardinality suggests that *Either* is similar to sum, *(,)* is similar to product and *(→)* to (flipped) power. These rules show that we can use many intuitions from high-school algebra when working with types.

Exercise 1.4. Counting *Maybes.* For each of the following types, enumerate and count the values:

1. *Bool → Maybe Bool*

2. *Maybe Bool → Bool*

3. *Maybe (Bool, Maybe (Bool, Maybe Bool))*

This is an opportunity to practice the learning outcome "develop adequate notation for mathematical concepts": what is a suitable notation for values of type *Bool, Maybe a, a → b, (a, b),* etc.? For those interested, Duregård et al. [2012] present a Haskell library (`testing-feat`) for enumerating syntax trees.

Exercise 1.5. Functions as tuples. For any type *t* the type *Bool → t* is basically "the same" as the type *(t, t).* Implement the two functions *isoR* and *isoL* forming an isomorphism:

$$isoR :: (Bool → t) → (t, t)$$
$$isoL :: (t, t) → (Bool → t)$$

and show that *isoL ∘ isoR = id* and *isoR ∘ isoL = id.*

Exercise 1.6. [**Important**] Functions and pairs (the "tupling transform"). From one function *f :: a → (b, c)* returning a pair, you can always make a pair of two functions *pf :: (a → b, a → c).* Implement this transform:

$$f2p :: (a → (b, c)) → (a → b, a → c)$$

Also implement the opposite transform:

$$p2f :: (a → b, a → c) → (a → (b, c))$$

This kind of transformation if often useful, and it works also for *n*-tuples.

Exercise 1.7. There is also a "dual" to the tupling transform: to show this, implement these functions:

$$s2p :: (Either\ b\ c → a) → (b → a, c → a)$$
$$p2s :: (b → a, c → a) → (Either\ b\ c → a)$$

Exercise 1.8. From Section 1.3:

- What does function composition do to a sequence? More concretely: for a sequence a what is $a \circ (1+)$? What is $(1+) \circ a$?

- How is $liftSeq_1$ related to $fmap$? $liftSeq_0$ to $conSeq$?

Exercise 1.9. Operator sections. Please fill out the remaining parts of this table with simplified expressions:

$$(1+) = \lambda x \to 1 + x$$
$$(*2) = \lambda x \to x * 2$$
$$(1+) \circ (*2) =$$
$$(*2) \circ (1+) =$$
$$= \lambda x \to x\hat{\ }2 + 1$$
$$= \lambda x \to (x+1)\hat{\ }2$$
$$(a+) \circ (b+) =$$

Exercise 1.10. Read the full chapter and complete the definition of the instance for *Num* for the datatype *ComplexSyn*. Also add a constructor for variables to enable writing expressions like $(Var \ "z") :*: toComplex \ 1$.

Exercise 1.11. We can embed semantic complex numbers in the syntax:

$embed :: ComplexSem \ r \to ComplexSyn \ r$
$embed \ (CS \ (x,y)) = ToComplexCart \ x \ y$

The embedding should satisfy a round-trip property: $eval \ (embed \ s)$ == s for all semantic complex numbers s. What about the opposite direction: when is $embed \ (eval \ e)$ == e? Here is a diagram showing how the types and the functions fit together

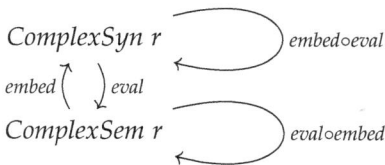

Step 0: type the quantification: what is the type of e?

Step 1: what equality is suitable for this type?

Step 2: if you use "equality up to eval" — how is the resulting property related to the first round-trip property?

Exercise 1.12. Read the next few pages of Appendix I (in [Adams and Essex, 2010]) defining the polar view of Complex Numbers and try to implement complex numbers again, this time based on magnitude and phase for the semantics. Try to use the same interface, so that you can import this new polar representation or the previous Cartesian representation without changing anything more than the import statement.

Exercise 1.13. Implement a simplifier $simp :: ComplexSyn\ r \rightarrow ComplexSyn\ r$ that handles a few cases like $0 * x = 0$, $1 * x = x$, $(a + b) * c = a * c + b * c$, etc. What class context do you need to add to the type of $simp$? This exercise is rather open-ended but it is recommended to start with a concrete view of what expected "normal forms" should be so that you can test it.

Exercise 1.14. A semiring is a set R equipped with two binary operations $+$ and $\cdot$, called addition and multiplication, such that:

- $(R, +, 0)$ is a commutative monoid with identity element 0:

$$(a + b) + c = a + (b + c)$$
$$a + b = b + a$$
$$0 + a = a + 0 = a$$

- $(R, \cdot, 1)$ is a monoid with identity element 1:

$$(a \cdot b) \cdot c = a \cdot (b \cdot c)$$
$$1 \cdot a = a \cdot 1 = a$$

- Multiplication left and right distributes over $(R, +, 0)$:

$$a \cdot (b + c) = (a \cdot b) + (a \cdot c)$$
$$(a + b) \cdot c = (a \cdot c) + (b \cdot c)$$
$$0 \cdot a = a \cdot 0 = 0$$

1. Define a datatype $SR\ v$ for the language of semiring expressions (with variables of type v). These are expressions formed from applying the semiring operations to the appropriate number of arguments, e.g., all the left hand sides and right hand sides of the above equations.

2. Implement the expressions from the laws.

3. Give a type signature for, and define, a general evaluator for $SR\ v$ expressions on the basis of an assignment function. An "assignment function" is a mapping from variable names to values.

Exercise 1.15. A *lattice* is a set L together with two operations $\vee$ and $\wedge$ (usually pronounced "sup" and "inf") such that

- $\vee$ and $\wedge$ are associative:

$$\forall\, x, y, z \in L.\ (x \vee y) \vee z = x \vee (y \vee z)$$
$$\forall\, x, y, z \in L.\ (x \wedge y) \wedge z = x \wedge (y \wedge z)$$

- $\vee$ and $\wedge$ are commutative:

$$\forall\, x, y \in L.\ x \vee y = y \vee x$$
$$\forall\, x, y \in L.\ x \wedge y = y \wedge x$$

- $\vee$ and $\wedge$ satisfy the *absorption laws*:

$$\forall\, x, y \in L.\ x \vee (x \wedge y) = x$$
$$\forall\, x, y \in L.\ x \wedge (x \vee y) = x$$

1. Define a datatype for the language of lattice expressions.

2. Define a general evaluator for *Lattice* expressions on the basis of an assignment function.

Exercise 1.16. An *abelian monoid* is a set M together with a constant (nullary operation) $0 \in M$ and a binary operation $\oplus : M \to M \to M$ such that:

- 0 is a unit of $\oplus$

$$\forall\, x \in M.\ 0 \oplus x = x \oplus 0 = x$$

- $\oplus$ is associative

$$\forall\, x, y, z \in M.\ x \oplus (y \oplus z) = (x \oplus y) \oplus z$$

- $\oplus$ is commutative

$$\forall\, x, y \in M.\ x \oplus y = y \oplus x$$

1. Define a datatype *AbMonoidExp* for the language of abelian monoid expressions. (These are expressions formed from applying the monoid operations to the appropriate number of arguments, e.g., all the left hand sides and right hand sides of the above equations.)

2. Define a general evaluator for *AbMonoidExp* expressions on the basis of an assignment function.

Chapter 2

DSLs for logic and proofs

In this chapter, we continue to exercise our skill of organising areas of mathematics in DSL terms. We apply our methodology to the languages of logic: propositions and proofs. Additionally, at the same time, we will develop adequate notions and notations for mathematical foundations and learn to perform calculational proofs. There will be a fair bit of theory: introducing propositional and first order logic, but also applications to mathematics: prime numbers, (ir)rationals, limit points, limits, etc. and some Haskell concepts.

2.1   Propositional Calculus

Our first DSL for this chapter is the language of *propositional calculus* (or propositional logic), modelling simple propositions with the usual combinators for and, or, implies, etc. When reading a logic book, one will encounter several concrete syntactic constructs related to propositional logic, which are collected in Table 2.1. Each row lists common synonyms and their arity.

False	$\perp$	F	nullary
True	$\top$	T	nullary
Not	$\neg$	$\sim$	unary
And	$\wedge$	&	binary
Or	$\vee$	$\mid$	binary
Implies	$\supset$	$\Rightarrow$	binary

Table 2.1: Syntactic constructors for propositions.

Some example propositions are $p_1 = a \wedge (\neg\, a)$, $p_2 = a \vee (\neg\, a)$, $p_3 = a \Rightarrow b$, $p_4 = (a \wedge b) \Rightarrow (b \wedge a)$. The names a, b, c, ... are "propositional variables": they can be substituted for any proposition. We could call them "variables", but in upcoming sections we will add another kind of variables (and quantification over them) to the calculus — so we keep calling them "names" to avoid confusing them.

Just as we did with complex number expressions in Section 1.5, we can model the abstract syntax of propositions as a datatype:

> **data** $Prop = Implies\ Prop\ Prop \mid And\quad Prop\ Prop \mid Or\quad Prop\ Prop$
> $\qquad\qquad\quad\ \mid Not\qquad Prop\qquad \mid Name\ Name\quad \mid Con\ Bool$
> **type** $Name = String$

The example expressions can then be expressed as

> $p_1, p_2, p_3, p_4 :: Prop$
> $p_1 = And\ (Name\ \texttt{"a"})\ (Not\ (Name\ \texttt{"a"}))$
> $p_2 = Or\quad (Name\ \texttt{"a"})\ (Not\ (Name\ \texttt{"a"}))$
> $p_3 = Implies\ (Name\ \texttt{"a"})\ (Name\ \texttt{"b"})$
> $p_4 = Implies\ (And\ a\ b)\quad (And\ b\ a)$
> $\quad$ **where** $a = Name\ \texttt{"a"}; b = Name\ \texttt{"b"}$

Because "names" stand for propositions, if we assign truth values for the names, we can compute a truth value of the whole proposition for the assignment in question.

2.1.1 An Evaluator for *Prop*

Let us formalise this idea in general, by writing an evaluator which takes a *Prop* to its truth value. (The evaluation function for a DSL describing a logic is often called *check* instead of *eval* but for consistency we stick to *eval*.)

> **type** $Env = Name \rightarrow Bool$
> $eval :: Prop \rightarrow Env \rightarrow Bool$
> $eval\ (Implies\ p\ q)\ env = eval\ p\ env \mathbin{\texttt{==>}} eval\ q\ env$
> $eval\ (And\ p\ q)\quad env = eval\ p\ env\ \wedge\ eval\ q\ env$
> $eval\ (Or\ p\ q)\quad\ env = eval\ p\ env\ \vee\ eval\ q\ env$
> $eval\ (Not\ p)\qquad env = \neg\ (eval\ p\ env)$
> $eval\ (Name\ n)\quad\ env = env\ n$
> $eval\ (Con\ t)\qquad env = t$
>
> $(\mathbin{\texttt{==>}}) :: Bool \rightarrow Bool \rightarrow Bool$
> $False \mathbin{\texttt{==>}} \_\quad = True$
> $True \mathbin{\texttt{==>}} p\quad = p$

F	$\Rightarrow$	a
F	T	F
F	T	T

a	$\Rightarrow$	b
F	T	F
F	T	T
T	F	F
T	T	T

a	$\wedge$	b	$\Rightarrow$	b	$\wedge$	a
F	F	F	T	F	F	F
F	F	T	T	T	F	F
T	F	F	T	F	F	T
T	T	T	T	T	T	T

(a) $t = F \Rightarrow a$ (b) $p_3 = a \Rightarrow b$.   (c) $p_4 = (a \wedge b) \Rightarrow (b \wedge a)$.

Figure 2.1: Truth table examples. Darker shades are filled in first, white column is the final result.

The function *eval* translates from the syntactic domain to the semantic domain, given an environment (an assignment of names to truth values), which we represent as a function from each *Name* to *Bool*. Here *Prop* is the (abstract) *syntax* of the language of propositional calculus and *Bool* is the *semantic domain*, and *env* : *Env* is a necessary extra parameter to write the function. Alternatively, and perhaps more elegantly, we can view *Env* → *Bool* as the semantic domain.

2.1.2 Truth tables and tautologies

Values of type *Name* → a are called "assignment functions" because they assign values (of type a) to the variable names. When we have $a = Bool$, and not too many variable names, we can collect all the combinations in a truth table.

As a first example of a truth table, consider the proposition $F \Rightarrow a$ which we call t here. The truth table semantics of t is usually drawn as in Fig. 2.1a: one column for each symbol, filled with the truth value of the expression "rooted" at that symbol. Thus, here we have one column for the name a listing all combinations of T and F, one (boring) column for F, and one column in the middle for the result of evaluating the expression. This table shows that no matter what value assignment we try for the only variable a, the semantic value is $T = True$. Thus the whole expression could be simplified to just T without changing its semantics.

If we continue with the example p_4 from above we have two names a and b which together can have any of four combinations of true and false. After the name-columns are filled, we fill in the rest of the table one operation (column) at a time (see Fig. 2.1c). The & columns become F F F T and finally the $\Rightarrow$ column (the output) becomes true everywhere. For our other examples, p_1 is always false, p_2 is always true, and p_3 is mixed.

A proposition whose truth table output is constantly true is called a *tautology*. Thus t, p_2 and p_4 are tautologies. We can formalise this idea as the following tautology-tester — a predicate which specifies the subset of *Prop*erties which are always true:

$$isTautology :: Prop \rightarrow Bool$$
$$isTautology\ p\ =\ and\ (map\ (eval\ p)\ (envs\ (freeNames\ p)))$$

It uses the helper functions *envs* to generate all possible environments (functions of type $Env = Name \rightarrow Bool$) for a given list of names and *freeNames* to find all names in a proposition. As an example, for p_4 above, *freeNames* would return the list $["a", "b"]$ and *envs* would return a four-element $[Env]$, one for each row in the truth table. The *map* would then apply *eval* p_4 to each element in the list to evaluate top-level truth value of the expression for each row. Finally *and* combines the results with $(\wedge)$ to ensure that they are all *True*.

$$envs :: [Name] \rightarrow [Env]$$
$$envs\ [\,]\qquad = [error\ "\texttt{envs: never used}"]$$
$$envs\ (n : ns) = [\ \ \lambda n' \rightarrow \textbf{if}\ n == n'\ \textbf{then}\ b\ \textbf{else}\ e\ n'$$
$$\qquad\qquad\qquad |\ b \leftarrow [False, True]$$
$$\qquad\qquad\quad,\ e \leftarrow envs\ ns$$
$$\qquad\qquad\quad]$$
$$freeNames :: Prop \rightarrow [Name]$$
$$freeNames = error\ "\texttt{exercise}"$$

Truth table verification is only viable for propositions with few names because of the exponential growth in the number of cases to check: for n names we get 2^n different rows in a truth table.

Exercise 2.1. Define the function *freeNames*.

There are much better algorithms to evaluate truth values than the naive one we just showed, but we will not go this route. Rather, we can introduce the notion of *proof*. (And in fact, the complexity of the best (known) algorithms for checking that a proposition is a tautology remain exponential in the number of variables.)

2.1.3 Proofs for Propositional Logic

Given a *Prop*osition p and a proof t (represented as an element of another type *Proof*), we can write a function that checks that t is a valid proof of p:

$$checkProof :: Proof \rightarrow Prop \rightarrow Bool$$

But we still have to figure out what consitutes proofs. We will build up the "proof DSL" one step at a time by looking at what we need to prove the different propositions.

To prove *And P Q*, one needs both a proof of *P* and a proof of *Q*. In logic texts, one will often find

$$\frac{P \quad Q}{P \wedge Q}$$

to represent this fact, which is called the *introduction rule for* ($\wedge$). For the proof to be complete, one still needs to provide a full proof of *P* and another for *Q* — it is not enough to just invoke this rule.

Therefore, in Haskell, we can represent this rule by a proof-term constructor *AndIntro* with two *Proof* arguments:

$$AndIntro :: Proof \rightarrow Proof \rightarrow Proof$$

and, the corresponding case of the *checkProof* function will look like this:

$$checkProof\ (AndIntro\ t\ u)\ (And\ p\ q) = checkProof\ t\ p \wedge checkProof\ u\ q$$

To prove *Or P Q*, we need either a proof of *P* or a proof of *Q* — but we need to know which side (*Left* for *p* or *Right* for *q*) we refer to. Therefore, we introduce two proof-term constructors:

$$OrIntroL :: Proof \rightarrow Proof$$
$$OrIntroR :: Proof \rightarrow Proof$$

There are a couple of possible approaches to deal with negation. One approach is to use De Morgan dualisation:

$$Not\ (a\ 'Or'\ b) = Not\ a\ 'And'\ Not\ b$$
$$Not\ (a\ 'And'\ b) = Not\ a\ 'Or'\ Not\ b$$
$$\ldots$$

Negation can then be pushed all the way down to names, which can recieve a special treatment in proof-checking.

However, we will instead apply the same treatment to negation as to other constructions, and define a suitable introduction rule:

$$\frac{P \rightarrow Q \quad P \rightarrow \neg Q}{\neg P}$$

(Intuitively, this rule says that to prove $\neg P$, one needs to derive a contradiction from *P*.) We can represent it by the constructor *NotIntro* :: *Prop* $\rightarrow$ *Proof* $\rightarrow$ *Proof* $\rightarrow$ *Proof*. Because we have inductive proofs (described from the bottom up), we have the additional difficulty that this rule conjures-up a new proposition, *Q*. This is why we need an additional *Prop* argument, which gives the *Q* formula.

There is no rule to introduce falsity ($\bot$) — otherwise we would have an inconsistent logic! Thus the last introduction rule deals with Truth, with no premise:

$$\frac{}{\top}$$

The proof has no information either: *TruthIntro* :: *Proof*.

To complete the system, in addition to introduction rules (where the connective appears as conclusion), we also need elimination rules (where the connective appears as premise). For conjunction (*And*), we have two eliminations rules:

$$\frac{P \wedge Q}{P} \quad \text{and} \quad \frac{P \wedge Q}{Q}$$

So we represent them by *AndElimL* :: *Prop* $\rightarrow$ *Proof* $\rightarrow$ *Proof* (and *AndElimR* symmetrically), where the extra *Prop* argument corresponds to *Q*.

For disjunction (*Or*) the idea is that if we know that $P \vee Q$ holds, then we have two cases: either *P* holds or *Q* holds. If only we can find a proposition *R* which is a consequence of both *P* and *Q*, then, regardless of which case we are facing, we know that *R* will hold. So, we get the following elimination rule for $P \vee Q$:

$$\frac{P \vee Q \quad P \rightarrow R \quad Q \rightarrow R}{R}$$

Our elimination for negation is $\frac{\neg\neg P}{P}$. It simply says that two negations cancel.

Finally we can eliminate falsity as follows:

$$\frac{\bot}{P}$$

This rule goes sometimes by its descriptive Latin name *ex falso quodlibet* — from falsehood, anything (follows).

We can then write our proof checker as follows. First the introduction rules:

$$
\begin{array}{lll}
checkProof\ TruthIntro & (Con\ True) & = True \\
checkProof\ (AndIntro\ t\ u) & (And\ p\ q) & = checkProof\ t\ p \\
& & \wedge\ checkProof\ u\ q \\
checkProof\ (OrIntroL\ t) & (Or\ p\ q) & = checkProof\ t\ p \\
checkProof\ (OrIntroR\ u) & (Or\ p\ q) & = checkProof\ u\ q \\
checkProof\ (NotIntro\ q\ t\ u) & (Not\ p) & = checkProof\ t\ (p\ \text{'Implies'}\ q) \\
& & \wedge\ checkProof\ u\ (p\ \text{'Implies'}\ Not\ q)
\end{array}
$$

then the elimination rules:

$$
\begin{aligned}
&checkProof\ (AndElimL\ q\ t) &&p = checkProof\ t\ (p\ 'And'\ q)\\
&checkProof\ (AndElimR\ p\ t) &&q = checkProof\ t\ (p\ 'And'\ q)\\
&checkProof\ (OrElim\ p\ q\ t\ u\ v)\ r &&= checkProof\ t\ (p\ 'Implies'\ r)\\
&&&\wedge\ checkProof\ u\ (q\ 'Implies'\ r)\\
&&&\wedge\ checkProof\ v\ (Or\ p\ q)\\
&checkProof\ (NotElim\ t) &&p = checkProof\ t\ (Not\ (Not\ p))\\
&checkProof\ (FalseElim\ t) &&p = checkProof\ t\ (Con\ False)
\end{aligned}
$$

Any other combination of proof/prop is an incorrect combination: the proof is not valid for the proposition.

$checkProof\ \_\ \_ = False$ -- incorrect proof

Once more, it can be interesting to view *checkProof* as an evaluator. This can be made plain by flipping its arguments: *flip checkProof* :: *Prop* $\to$ (*Proof* $\to$ *Bool*). This way, one can understand *Proof* $\to$ *Bool*, a subset of proofs, as the semantic domain of *Prop*. In other words, a proposition can be interpreted as the subset of proofs which prove it.

2.1.4 Implication, hypothetical derivations, contexts

We have so far omitted to deal with *Implies*. One reason is that we can use the so-called material implication definition which we invoked earlier in truth tables. It means to define *Implies a b* = (*Not a*) '*Or*' *b* — and this equality means that there is no need to deal specially with *Implies*. However this approach does not bring any new insight. In particular, this view is hard to transport to more complicated logics (such as second-order logic).

Thus we take our usual approach and give rules for it. The introduction rule is sometimes written in this way in logic texts:

$$
\begin{array}{c}
P\\
\vdots\\
\dfrac{Q}{P \to Q}
\end{array}
$$

Such a notation can, however, be terribly confusing. We were already used to the fact that proofs above the line had to be continued, so what can the dots possibly mean? The intended meaning of this notation is that, to prove $P \to Q$, it suffices to prove Q, but one is also allowed to use P as an assumption in this (local) proof of Q.

By adding a constructor corresponding to implication introduction: $ImplyIntro ::$ $(Proof \rightarrow Proof) \rightarrow Proof$, we can formalise this rule as in the Haskell data type representing our proof DSL. The fact that the premise can depend on the assumption Q is represented by a function whose parameter is the proof of Q in question. In other words, to prove the formula $P \rightarrow Q$ we assume a proof t of P and derive a proof u of Q. So, a proof of an implication is a function from proofs to proofs.

The eliminator for implication (also known as *modus ponens*) is

$$\frac{P \rightarrow Q \quad P}{Q}$$

We formalise it as $ImplyElim :: Prop \rightarrow Proof \rightarrow Proof \rightarrow Proof$[1]. And we can finally complete our proof checker as follows:

```
checkProof (Assume p')        p              = p == p'
checkProof (ImplyIntro f)     (p 'Implies' q) = checkProof (f (Assume p)) q
checkProof (ImplyElim p t u)  q              = checkProof t (p 'Implies' q)
                                               ∧ checkProof u p
checkProof _                  _              = False   -- incorrect proof
```

And, for reference, the complete DSL for proofs is given by the following datatype:

```
data Proof = TruthIntro              | FalseElim Proof
           | AndIntro   Proof Proof
           | AndElimL Prop  Proof    | AndElimR Prop Proof
           | OrIntroL   Proof        | OrIntroR   Proof
           | OrElim Prop Prop Proof Proof Proof
           | NotIntro Prop Proof Proof  | NotElim Proof
           | Assume Prop
           | ImplyIntro (Proof → Proof) | ImplyElim Prop Proof Proof
```

Aside   The *Assume* constructor may make the reader somewhat uneasy: how come that we can simply assume anything? The intent is that this constructor is *private* to the *checkProof* function (or module). No user-defined proof can use it. The most worried readers can also define the following version of *checkProof*, which uses an extra context to check that assumption have been rightfully introduced earlier.[2]

---

[1]The proposition P is not given by the conclusion and thus is provided as part of the proof.

[2]For the *cognoscenti*, this kind of presentation of the checker matches well the sequent calculus presentation of the proof system

> *checkProof'* :: *Context* → *Proof* → *Prop* → *Bool*
> *checkProof' ctx* (*ImplyIntro t*) (*p* '*Implies*' *q*) = *checkProof'* (*p* : *ctx*) *t q*
> *checkProof' ctx* (*Assume* _) *p* = *p* ∈ *ctx*

Example proof We can put our proof-checker to the test by writing some proofs and verifying them.

> *conjunctionComm* :: *Prop*
> *conjunctionComm* = p_4
>
> *conjunctionCommProof* :: *Proof*
> *conjunctionCommProof* = *ImplyIntro step*
>   **where** *step* :: *Proof* → *Proof*
>   *step evAB* = *AndIntro* (*AndElimR* (*Name* "a") *evAB*)
>   (*AndElimL* (*Name* "b") *evAB*)

where *evAB* stands for "evidence for *A* and *B*". We can then run the checker and verify: *checkProof conjunctionCommProof conjunctionComm* == *True*

Exercise 2.2. Try to swap *AndElimL* and *AndElimR* in the above proof. What will happen and why?

Or **is the dual of** *And* Before moving on to our next topic, we make a final remark on *And* and *Or*. Most of the properties of *And* have corresponding properties for *Or*. This can be explained one way by observing that they are De Morgan duals. Another explanation is that one can swap the direction of the arrows in the types of the role between introduction and elimination. (Using our presentation, doing so requires applying isomorphisms.)

2.1.5 The Haskell type-checker as a proof checker

Perhaps surprisingly, the proof-checker that we just wrote is already built-in in the Haskell compiler. Let us clarify what we mean, using the same example, but adapt it to let the type-checker do the work:

> *conjunctionCommProof'* :: *Implies* (*And a b*) (*And b a*)
> *conjunctionCommProof'* = *implyIntro step*
>   **where** *step* :: *And a b* → *And b a*
>   *step evAB* = *andIntro* (*andElimR evAB*)
>   (*andElimL evAB*)

That is, instead of writing propositions, we write types (*And, Or, Implies* — which we leave abstract for now). Instead of using *Proof* constructors, we use functions whose types capture rules:

```
truthIntro :: Truth
falseElim  :: False → p
andIntro   :: p → q → And p q
andElimL  :: And p q → p
andElimR  :: And p q → q
orIntroL  :: p → Or p q
orIntroR  :: q → Or p q
orElim     :: Or p q → (p 'Implies' r) → (q 'Implies' r) → r
notIntro   :: (p 'Implies' q) 'And' (p 'Implies' Not q) → Not p
notElim    :: Not (Not p) → p
implyIntro :: (p → q) → (p 'Implies' q)
implyElim :: (p 'Implies' q) → (p → q)
```

Instead of running *checkProof*, we type-check the above program. Because the proof is correct, we get no type-error.

Exercise 2.3. What would happen if you swap *andElimR* and *andElimL*? Why?

This style of propositional logic proof is very economical, because not only the checker comes for free, but we additionally get all the engineering tools of the Haskell tool-chain.

One should be careful however that Haskell is not designed with theorem-proving in mind. For this reason it is easily possible make the compiler accept invalid proofs. The main two sources of invalid proofs are 1. non-terminating programs and 2. exception-raising programs. In sum, the issue is that Haskell allows the programmer to define partial functions (instead of total ones, see Section 1.1.3).

2.1.6   Intuitionistic Propositional Logic

We can make the link between Haskell and logic more tight if we restrict ourselves to *intuitionistic* logic.

One way to characterise intuitionistic logic is that it lacks native support for negation. Instead, *Not p* is represented as *p 'Implies' False*:

type *Not p = p 'Implies' False*

The intuition behind this definition is the principle of proof by contradiction: if assuming p leads to a contradiction (*False*), then p must be false; so *Not p* should hold.

When doing this kind of definition, one gives up on *notElim*: there is no way to eliminate (double) negation.

> *notElim* = *error* "not possible as such in intuitionistic logic"

On the other hand the introduction rule for negation becomes a theorem of the logic. The formulation of the theorem is:

> *notIntro* :: $(p\,'Implies'\,q)\,'And'\,(p\,'Implies'\,Not\,q) \rightarrow Not\,p$

where *Not p* = $p\,'Implies'\,False$, and we give its proof below, after introducing concrete representations of implications and conjunction.

Representing *Implies* **and** *And* By focusing on intuitionistic logic, we can give a *typed* representation for each of the formula constructors. Let us consider implication first. The proof rules *impIntro* and *impElim* seem to be conversion from and to functions, and so it should be clear that the representation of the implication formula is a function:

> **type** *Implies p q* = $p \rightarrow q$
> *implyElim f* = *f*
> *implyIntro f* = *f*

Conjunction is represented as pairs; that is, if $t : P$ and $u : Q$ then the proof of *And P Q* should be a pair (t, u). The elimination rules are projections. In code:

> **type** *And p q* = (p, q)
> *andIntro t u* = (t, u)
> *andElimL* = *fst*
> *andElimR* = *snd*

Proof of *notIntro* Before introducing the representations of the remaining constructors, let us get back to the proof of *notIntro*. If we introduce abbreviations and simplify the types using our newly introduced representations we get:

> *P2Q* = $p\,'Implies'\,q$ = $p \rightarrow q$
> *P2nQ* = $p\,'Implies'\,Not\,q = p \rightarrow Not\,q = p \rightarrow q \rightarrow False$

And as we know *And* is just the pair type constructor, the type of *notIntro* is

$$And\ P2Q\ P2nQ \rightarrow Not\ p = (P2Q, P2nQ) \rightarrow p \rightarrow False$$

Thus we can start the definition by matching on a pair and a proof:

$$notIntro\ (evPimpQ, evPimpNotQ)\ evP =$$
      -- a proof of *False* wanted

where *evPimpQ* :: *P2Q* and *evPimpNotQ* :: *P2nQ* are functions from *p*. We can
now apply these two functions to *evP* :: *p* and continue the body of *notIntro*:

let *evQ*   = *evPimpQ*   *evP*
      evNotQ = *evPimpNotQ* *evP*

to give us both a proof *evQ* :: *q* and a function *evNotQ* :: *q* $\rightarrow$ *False*. Finally, we
just apply this function to the proof:

in *evNotQ evQ*

If we inline the local definitions and give the arguments shorter names we get
a one-liner:

$$notIntro :: (p \rightarrow q, p \rightarrow q \rightarrow False) \rightarrow p \rightarrow False$$
$$notIntro\ (f, g)\ x = (g\ x)\ (f\ x)$$

More importantly, we can see that the types help by, in effect, forcing the definition to be (equivalent to) this one.

Back to representations   Disjunction is represented as *Either*: if $p : P$ then
Left p : Or P Q and if $q : Q$ then *Right q : Or P Q*.

type *Or a b = Either a b*
orIntroL = Left
orIntroR = Right
orElim pOrq f g = **case** *pOrq* **of**
   Left   p $\rightarrow$ *f p*
   Right q $\rightarrow$ *g q*

We already had characterised or-elimination as case analysis, and, indeed, this
is how we implement it.

Truth is represented as the unit type:

type *Truth* = ()
truthIntro = ()

And falsehood is represented as the *empty* type (with no constructor):

> **data** *False*
> *falseElim x* = **case** *x* **of** { }

Note that the case-analysis has nothing to take care of here.

In this way we can build proofs ("proof terms") for all of intuitionistic propositional logic (IPL). As we have seen, each such proof term is a program in Haskell. Conversely, every program written in this fragment of Haskell (functions, pairs, *Either*, no recursion and full coverage of cases) can be turned into a proof in IPL. This fragment is called the simply-typed lambda calculus (STLC) with sum and products.

2.1.7 Type-Driven Development of Proofs as Programs

With the logic connectives implemented as type constructors we explore a few more examples of laws and their proofs.

The law of the excluded middle   As an example of how intuitionism affects logic, consider the law of the excluded middle, which states that, for any proposition *P*, either *P* or *Not P* holds. For example, either it rains or it does not rain. There is no "middle ground". If we attempt to prove *Or P* (*Not P*) in intuitionistic logic, we quickly find ourselves in a dead end. Clearly, we cannot prove *P* for any *P*. Likewise *Not P*, or equivalently $P \rightarrow False$ cannot be deduced.

What we have to do is to account for the fact that we cannot use negation elimination, and so we have to make-do with proving *Not* (*Not Q*) instead of *Q*. This is exactly what we have to do to (almost) prove the law of excluded middle. First we expand the type using the definitions of *Not* and *Or*:

> *Not* (*Not* (*p 'Or' Not p*))
> = {- Def. of the outermost *Not* -}
> *Not* (*p 'Or' Not p*) $\rightarrow$ *False*
> = {- Def. of *Not* on the left -}
> ((*p 'Or' Not p*) $\rightarrow$ *False*) $\rightarrow$ *False*
> = {- Def. of *Or* -}
> (*Either p* (*Not p*) $\rightarrow$ *False*) $\rightarrow$ *False*
> = {- Def. of the last *Not* -}
> (*Either p* (*p* $\rightarrow$ *False*) $\rightarrow$ *False*) $\rightarrow$ *False*

Thus the proof needs to be a rather unusual higher-order function. The first argument, which we can call *k*, is also a function; from *Either p* (*p* $\rightarrow$ *False*)

to *False*. We can use this function in two ways: apply it to *Left evP* if we
have a proof *evP* :: *p* around or to *Right evNotP* if we can build a function
evNotP :: *p* → *False*. To start out we cannot take the *Left* path, so we must
try the *Right* path. Then, what is left is the (local) definition of *evNotP*. This
function will receive as an argument a value *evP* :: *p*, and now we can take the
Left path, using *k* again.

In code with comments we get this Haskell-encoded proof:

> *excludedMiddle* :: *Not* (*Not* (*p* '*Or*' *Not p*))   -- to prove this, we can ...
> *excludedMiddle k* =   -- ... assume *Not* (*Or p* (*Not p*)) and prove falsity.
>   *k*   -- So, we can prove falsity if we can prove *Or p* (*Not p*).
>   (*Right*   -- We can prove in particular the right case, *Not p*
>   (λ*evP* →   -- ... by assuming that *p* holds, and prove falsity.
>   *k*   -- Again, we can prove falsity if we can prove *Or p* (*Not p*).
>   (*Left*   -- This time, we can prove in particular the left case, *p*
>   *evP*)))   -- because we assumed it earlier!

which can be shortened to a one-liner if we cut out the comments:

> *excludedMiddle'* :: *Not* (*Not* (*p* '*Or*' *Not p*))
> *excludedMiddle' k* = *k* (*Right* (λ*evP* → *k* (*Left evP*)))

Revisiting the tupling transform In Exercise 1.6, the "tupling transform"
was introduced, relating a pair of functions to a function returning a pair.
(Revisit that exercise if you skipped it before.) There is a logic formula corresponding to the type of the tupling transform:

> (*a* '*Implies*' (*b* '*And*' *c*)) '*Iff*' (*a* '*Implies*' *b*) '*And*' (*a* '*Implies*' *c*)

(*Iff* refers to implication in both directions). The proof of this formula closely
follows the implementation of the transform. Therefore we start with the two
directions of the transform as functions:

> *test1'* :: (*a* → (*b*, *c*)) → (*a* → *b*, *a* → *c*)
> *test1' a2bc* = (λ*a* → *fst* (*a2bc a*)
>   , λ*a* → *snd* (*a2bc a*))
> *test2'* :: (*a* → *b*, *a* → *c*) → (*a* → (*b*, *c*))
> *test2' fg* = λ*a* → (*fst fg a*, *snd fg a*)

Then we move on to the corresponding logic statements with proofs. Note
how the functions are "hidden inside" the proof.

$test1 ::$ *Implies* (*Implies a* (*And b c*)) (*And* (*Implies a b*) (*Implies a c*))
$test1 =$ *implyIntro* ($\lambda a2bc \rightarrow$
        andIntro (*implyIntro* ($\lambda a \rightarrow andElimL$ (*implyElim a2bc a*)))
            (*implyIntro* ($\lambda a \rightarrow andElimR$ (*implyElim a2bc a*))))
$test2 ::$ *Implies* (*And* (*Implies a b*) (*Implies a c*)) (*Implies a* (*And b c*))
$test2 =$ *implyIntro* ($\lambda fg \rightarrow$
        implyIntro ($\lambda a \rightarrow$
            andIntro (*implyElim* (*andElimL fg*) a)
                (*implyElim* (*andElimR fg*) a)))

Logic as impoverished typing rules Another view of the same isomorphism is that the logical rules for IPL can be obtained by erasing programs from the typing rules for STLC. We will show here only the application rule, leaving the rest as an exercise. This typing rule for function application can be written as follows:

$$\frac{f : A \rightarrow B \quad x : A}{f(x) : B}$$

After erasing the colon (:) sign and what comes before it, we obtain *modus ponens* — implication elimination.

The *Curry–Howard correspondence* is a general principle that says that we can think of propositions as types, and proofs as programs. This principle goes beyond propositional logic (and even first order logic): it applies to all sorts of logics and programming languages, with various levels of expressivity and features.

2.2 First Order Logic

module *DSLsofMath.FOL* **where**

Our next DSL is that of *First Order Logic*, or FOL for short, and also known as Predicate Logic. Compared to propositional logic, the main addition is *quantification over individuals*. Additionally, one adds a language of terms — its semantic domain being the individuals which we quantify over.

Let us study terms first. A *term* is either a (term) *variable* (like x, y, z), or the application of a *function symbol* (like f, g) to a suitable number of terms. If we have the function symbols f of arity 2 and g of arity 3 we can form terms like

$f(x,x), g(y,z,z), g(x,y,f(x,y))$, etc. The individuals are often limited to a single domain. For example here we will take individuals to be rationals (to be able to express basic mathematical concepts). Consequently, the actual function symbols are also domain-specific — for rationals we will have addition, division, etc. In this case we can model the terms as a datatype:

> **type** *VarT* = *String*
> **data** *RatT* = *RV VarT* | *FromI Integer* | *RPlus RatT RatT* | *RDiv RatT RatT*
>   **deriving** *Show*

The above introduces variables (with the constructor *RV*) and three function symbols: *FromI* of arity 1, *RPlus*, *RDiv* of arity 2.

Exercise 2.4. Following the usual pattern, write the evaluator for *RatT*:

> *evalRat* :: *RatT* → (*VarT* → *RatSem*) → *RatSem*
> **type** *RatSem* = *Rational*

As mentioned above, the propositions (often referred to as *formulas* in the context of FOL) are extended so that they can refer to terms. That is, the names from the propositional calculus are generalised to *predicate symbols* of different arity. The predicate symbols can only be applied to terms, not to other predicate symbols or formulas. As an arbitrary example, if we have the predicate symbols *Positive* of arity 1 and *LessThan* of arity 2 we can form *formulas* like *Positive* (x), *LessThan* $(f (x,x), y)$, etc. Note that we have two separate layers, with terms at the bottom: formulas normally refer to terms, but terms cannot refer to formulas.

The formulas introduced so far are all *atomic formulas*: generalisations of the *names* from *Prop*. Now we will add two more concepts: first the logical connectives from the propositional calculus: *And, Or, Implies, Not*, and then two quantifiers: "forall" ($\forall$) and "exists" ($\exists$). Thus the following are examples of FOL formulas:

> $\forall x.$ *Positive* $(x) \Rightarrow (\exists y.$ *LessThan* $(f (x,x), y))$
> $\forall x.$ $\forall y.$ *Equal* $(plus (x,y), plus (y,x))$ -- *plus* is commutative

Here is the second formula again, but with infix operators:

> $\forall x.$ $\forall y.$ $(x + y)$ == $(y + x)$

Note that (==) is a binary predicate symbol (written *Equal* above), while (+) is a binary function symbol (written *plus* above).

The fact that quantification is over individuals is a defining characteristic of FOL. If one were to, say, quantify over predicates, we would have a higher-order logic, with completely different properties.

As before we can model the expression syntax (for FOL, in this case) as a datatype. We keep on using the logical connectives *Implies, And, Or, Not* from the type *Prop*, add predicates over terms, and quantification. The constructor *Equal* could be eliminated in favour of *PName* "Equal" but it is often included as a separate constructor.

> **type** *PSym* $=$ *String*
> **data** *FOL* $=$ *Implies FOL FOL* | *And FOL FOL* | *Or FOL FOL* | *Not FOL*
> | *FORALL VarT FOL* | *EXISTS VarT FOL*
> | *PName PSym* $[RatT]$ | *Equal RatT RatT*
>   **deriving** *Show*

2.2.1   Evaluator for Formulas and *Undecidability

Setting us up for failure, let us attempt to write an *eval* function for FOL, as we did for propositional logic.

In propositional logic, we allowed the interpretation of propositional variables to change depending on the environment. Here, we will let the interpretation of term variables be dependent on an environment, which will therefore map (term) variables to individuals (*VarT* $\rightarrow$ *RatSem*). If we so wished, we could have an environment for the interpretation of predicate names, with an environment of type *PSym* $\rightarrow$ $[RatSem]$ $\rightarrow$ *Bool*. Rather, with little loss of generality, we will fix this interpretation, via a constant function *eval0*, which may look like this:

> *eval0* :: *PSym* $\rightarrow$ $[RatSem]$ $\rightarrow$ *Bool*
> *eval0* "Equal" $[t_1, t_2]$ $= t_1$ == t_2
> *eval0* "LessThan" $[t_1, t_2]$ $= t_1 < t_2$
> *eval0* "Positive" $[t_1]$ $= t_1 > 0$

So we would use the following type, and go our merry way for most cases:

> *eval* :: *FOL* $\rightarrow$ (*VarT* $\rightarrow$ *RatSem*) $\rightarrow$ *Bool*
> *eval formula env* $=$ *ev formula*
>   **where** *ev* (*PName n args*) $=$ *eval0 n* (*map* (*flip evalRat env*) *args*)
>   *ev* (*Equal a b*) $=$ *evalRat a env* == *evalRat b env*
>   *ev* (*And p q*) $=$ *ev p* $\wedge$ *ev q*
>   *ev* (*Or p q*) $=$ *ev p* $\vee$ *ev q*

However, as soon as we encounter quantifiers, we have a problem. To evaluate *EXISTS* x p (at least in certain contexts) we may need to evaluate p for each possible value of x. But, unfortunately, there are infinitely many such possible values, and so we can never know if the formula is a tautology.[3] So, if we were to try and run the evaluator, it would not terminate. Hence, the best that we can ever do is, given a hand-written proof of the formula, check if the proof is valid. Fortunately, we have already studied the notion of proof in the section on propositional logic, and it can be extended to support quantifiers.

2.2.2 Universal quantification

Universal quantification (Forall or $\forall$) can be seen as a generalisation of *And*. To see this, we can begin by generalising the binary operator *And* to an *n*-ary version: And_n. To prove And_n $(A_1, A_2, \ldots, A_n)$ we need a proof of each A_i. Thus we could define And_n $(A_1, A_2, \ldots, A_n) = A_1 \& A_2 \& \ldots \& A_n$ where $\&$ is the infix version of binary *And*. The next step is to require the formulas A_i to be of the same form, i.e. the result of applying a constant function A to the individual i. And, we can think of the variable i ranging over the full set of individuals $i_1, i_2, \ldots$. Then the final step is to introduce the notation $\forall i.$ A (i) for A $(i_1) \& A$ $(i_2) \& \ldots$.

Now, a proof of $\forall x.$ A (x) should in some way contain a proof of A (x) for every possible x. For the binary *And* we simply provide the two proofs, but in the infinite case, we need an infinite collection of proofs. To do so, a possible procedure is to introduce a fresh (meaning that we know nothing about this new term) constant term a and prove A (a). Intuitively, if we can show A (a) without knowing anything about a, we have proved $\forall x.$ A (x). Another way to view this is to say that a proof of $\forall x.$ P x is a function f from individuals to proofs such that f t is a proof of P t for each term t.

So we can now extend our type for proofs: the introduction rule for universal quantification is $\frac{A(x) \quad x \text{ fresh}}{\forall x. A(x)}$. The corresponding constructor can be $\forall$-*Intro* :: $(RatSem \rightarrow Proof) \rightarrow Proof$. (In Haskell we would use the textual syntax *AllIntro* for $\forall$-*Intro*) and *AllElim* for $\forall$-*Elim*.)

Note that the scoping rule for $\forall x.$ b is similar to the rule for a function definition, f $x = b$, and for anonymous functions, $\lambda x \rightarrow b$. Just as in those cases we

---

[3]FOL experts will scoff at this view, because they routinely use much more sophisticated methods of evaluation, which handle quantifiers in completely different ways. Their methods are even able to identify tautologies as such. However, even such methods are not guaranteed to terminate on formulas which are not tautologies. Therefore, as long as an even-very-advanced FOL tautology-checker is running, there is no way to know how close it is to confirming if the formula at hand is a tautology or not. This is not a technical limitation, but rather a fundamental one, which boils down to the presence of quantifiers over an infinite domain.

say that the variable x is *bound* in b and that the *scope* of the variable binding extends until the end of b (but not further).

One common source of confusion in mathematical (and other semi-formal) texts is that variable binding is sometimes implicit. A typical example is the notation for equations: for instance $x^2 + 2 * x + 1 == 0$ usually means roughly $\exists x.\ x^2 + 2 * x + 1 == 0$. We write "roughly" here because in maths texts the scope of x very often extends to some text after the equation where something more is said about the solution x.[4]

Let us now consider the elimination rule for universal quantification. The idea here is that if $A\ (x)$ holds for every abstract individual x, then it also holds for any concrete individual a: $\frac{\forall x.A(x)}{A(a)}$. As for *And* we had to provide the other argument to recover $p\ 'And'\ q$, here we have to be able reconstruct the general form $A\ (x)$ — indeed, it is not simply a matter of substituting x for a, because there can be several occurrences of a in the formula to prove. So, in fact, the proof constructor must contain the general form $A\ (x)$, for example as a function from individuals: $\forall\text{-}Elim :: (RatSem \rightarrow Prop) \rightarrow Proof \rightarrow Proof$.

Let us sketch the proof-checker cases corresponding to universal quantification. The introduction rule uses a new concept: *subst x a p*, which replaces the variable x by a in p, but otherwise follows closely our informal explanation:

$$proofChecker\ (\forall\text{-}Intro\ f\ a)\ (\forall\ x.\ p) = proofChecker\ (f\ a')\ (subst\ x\ a'\ p)$$
$$\textbf{where}\ a' = freshFor\ [a, \forall\ x.\ p]$$
$$proofChecker\ (\forall\text{-}Elim\ f\ t)\ p = checkUnify\ (f\ x')\ p$$
$$\wedge\ proofChecker\ t\ (\forall\ x'.\ f\ x')$$
$$\textbf{where}\ x' = freshFor\ [f\ x, p]$$

The eliminator uses *checkUnify* which verifies that $f\ x$ is indeed a generalisation of the formula to prove, p. Finally we need a way to introduce fresh variables *freshFor*, which conjures up a variable occurring nowhere in its argument list.

2.2.3 Existential quantification

We have already seen how the universal quantifier can be seen as a generalisation of *And* and in the same way we can see the existential ($\exists$) quantifier as a generalisation of *Or*.

First we generalise the binary *Or* to an *n*-ary Or_n. To prove $Or_n\ A_1\ A_2 \ldots A_n$ it is enough (and necessary) to find one i for which we can prove A_i. As before

---

[4]This phenomena seems to be borrowed from the behaviour of quantifiers in natural language. See for example [Bernardy et al., 2021] for a discussion.

we then take the step from a family of formulas A_i to a single unary predicate A expressing the formulas A (i) for the (term) variable i. Then the final step is to take the disjunction of this infinite set of formulas to obtain $\exists\, i.\ A\ i$.

The elimination and introduction rules for existential quantification are:

$$\frac{P(a)}{\exists i.P(i)} \qquad\qquad \frac{\forall x.P(x) \to R \qquad \exists i.P(i)}{R}$$

The introduction rule says that to prove the existential quantification, we only need exhibit one witness (a) and one proof for that member of the set of individuals. For binary *Or* the "family" only had two members, one labelled *Left* and one *Right*, and we effectively had one introduction rule for each. Here, for the generalisation of *Or*, we have unified the two rules into one with an added parameter a corresponding to the label which indicates the family member.

In the other direction, if we look at the binary elimination rule, we see the need for two arguments to be sure of how to prove the implication for any family member of the binary *Or*.

$$orElim :: Or\ p\ q \to (p\ 'Implies'\ r) \to (q\ 'Implies'\ r) \to r$$

The generalisation unifies these two to one family of arguments. If we can prove R for each member of the family, we can be sure to prove R when we encounter some family member.

The constructors for proofs can be $\exists Intro :: RatSem \to Proof \to Proof$ and $\exists Elim :: Proof \to Proof \to Proof$. In this case we would have i as the first argument of $\exists Intro$ and a proof of A (i) as its second argument.

Exercise 2.5. Sketch the *proofChecker* cases for existential quantification.

2.2.4   Typed quantification

So far, we have considered quantification always as over the full set of individuals, but it is often convenient to quantify over a subset with a certain property (like all even numbers, or all non-empty sets).

Even though it is not usually considered as strictly part of FOL, it does not fundamentally change its character if we extended it with several types (or sorts) of individuals (one speaks of "multi-sorted" FOL).

In such a variant, the quantifiers look like $\forall\, x : S.\ P\ x$ and $\exists\, x : S.\ P\ x$.

Indeed, if a type (a set) S of terms can be described as those that satisfy the unary predicate T we can understand $\forall\, x : T.\ P\ x$ as a shorthand for $\forall\, x.\ T\ x \Rightarrow P\ x$. Likewise we can understand $\exists\, x : T.\ P\ x$ as a shorthand for $\exists\, x.\ T\ x\ \&\ P\ x$.

As hinted at in the previous chapters, we find that writing types explicitly can greatly help understanding, and we won't refrain from writing down types in quantifiers in FOL formulas.

Exercise 2.6. Prove that the De Morgan dual of typed universal quantification is the typed existential quantification, using the above translation to untyped quantification.

2.2.5   Curry-Howard for quantification over individuals

We can try and draw parallels with a hypothetical programming language corresponding to FOL. In such a programming language, we expect to be able to encode proof rules as follows (we must use a *dependent* function type here, $(a : A) \to B$, see below)

$$
\begin{array}{ll}
\textit{allIntro} & :: ((a : \textit{Individual}) \to P\, a) \to (\forall\, x.\ P\, x) \\
\textit{allElim} & :: (\forall\, x.\ P\, x) \to ((a : \textit{Individual}) \to P\, a) \\
\textit{existIntro} & :: (a : \textit{Individual}) \to P\, a \to \exists\, x.\ P\, x \\
\textit{existsElim} & :: ((a : \textit{Individual}) \to P\, a\, \text{'Implies'}\, R) \to (\exists\, x.\ P\, x)\, \text{'Implies'}\, R
\end{array}
$$

Taking the intuitionistic version of FOL (with the same treatment of negation as for propositional logic), we additionally expect to be able to represent proofs of quantifiers, directly. That is:

(t, b_t) is a program of type $\exists\, x.\ P\, x$ if b_t is has type $P\, t$.

f is a program of type $\forall\, x.\ P\, x$ if $f\, t$ is has type $P\, t$ for all t.

Unfortunately, in its 2010 standard, Haskell does not provide the equivalent of quantification over individuals. Therefore, one would have to use a different tool than Haskell as a proof assistant for (intuitionistic) FOL. The quantification that Haskell provides (*forall a. ...*) is *over types* rather than individuals.What we would need is: 1. a type corresponding to universal quantification, the dependent function type $(a : A) \to B$, and 2. a type corresponding to $\exists\, x : A.\ P\, x$, the dependent pair $(x : A, P\, x)$.

We can recommend the language Agda [Norell, 2009] or Idris [Brady, 2016] if you want to express these directly in the language (as done in [Mu et al., 2008, 2009, Ionescu and Jansson, 2012, Botta et al., 2017a]). However, in order to avoid a multiplicity of tools and potentially an excessive emphasis on proof formalism, we will refrain from formalising proofs as Agda or Idris programs in the remainder. Rather, in the rest of the chapter, we will illustrate the logical principles seen so far by examples.

2.3   An aside: Pure set theory

One way to build mathematics from the ground up is to start from pure set theory and define all concepts by translation to sets. We will only work with (a small corner of) this as a mathematical domain to study, not as "the right way" of doing mathematics (there are other ways). To classify the sets we will often talk about the *cardinality* of a set which is defined as the number of elements in it.

The core of the language of pure set theory is captured by four function symbols ({ }, S, *Union*, and *Intersection*). We use a nullary function symbol { } to denote the empty set (sometimes written $\varnothing$) and a unary function symbol S for the function that builds a singleton set from an "element". All non-variable terms so far are { }, S { }, S (S { }), ... The first set is empty but all the others denote (different) one-element sets.

Next we add two binary function symbols for union and intersection of sets (denoted by terms). Using union we can build sets of more than one element, for example *Union* (S { }) (S (S { })) which has two "elements": { } and S { }.

In pure set theory we don't actually have any distinguished "elements" to start from (other than sets), but it turns out that quite a large part of mathematics can still be expressed. Every term in pure set theory denotes a set, and the elements of each set are again sets.

At this point it is a good exercise to enumerate a few sets of cardinality 0, 1, 2, and 3. There is really just one set of cardinality 0: the empty set s_0 = { }. Using S we can then construct $s_1 = S\ s_0$ of cardinality 1. Continuing in this manner we can build $s_2 = S\ s_1$, also of cardinality 1, and so on. Now we can combine different sets (like s_1 and s_2) with *Union* to build sets of cardinality 2: $s_3 =$ *Union* $s_1\ s_2$, $s_4 =$ *Union* $s_2\ s_3$, etc. And we can at any point apply S to get back a new set of cardinality 1, like $s_5 = S\ s_3$.

Natural numbers To talk about things like natural numbers in pure set theory they need to be encoded. FOL does not have function definitions or recursion, but in a suitable meta-language (like Haskell) we can write a function that creates a set with n elements (for any natural number n) as a term in FOL. Here is some pseudocode defining the "von Neumann" encoding:

```
vN 0      = { }
vN (n + 1) = step (vN n)
step x = Union x (S x)
```

If we use conventional set notation we get $vN\ 0$ = { }, $vN\ 1$ = {{ }}, $vN\ 2$ = {{ }, {{ }}}, $vN\ 3$ = {{ }, {{ }}, {{ }, {{ }}}}, etc. If we use the shorthand

$\bar{n}$ for vN n we see that $\bar{0} = \{\,\}$, $\bar{1} = \{\bar{0}\}$, $\bar{2} = \{\bar{0},\bar{1}\}$, $\bar{3} = \{\bar{0},\bar{1},\bar{2}\}$ and, in general, that $\bar{n}$ has cardinality n (meaning it has n elements).

Pairs The constructions presented so far show that, even starting from no elements, we can embed all natural numbers in pure set theory. We can also embed unordered pairs: $\{a,b\} \stackrel{\text{def}}{=} Union\ (S\ a)\ (S\ b)$ and normal, ordered pairs: $(a,b) \stackrel{\text{def}}{=} \{S\ a, \{a,b\}\}$. With a bit more machinery it is possible to step by step encode $\mathbb{N}$, $\mathbb{Z}$, $\mathbb{Q}$, $\mathbb{R}$, and $\mathbb{C}$. A good read in this direction is "The Haskell Road to Logic, Maths and Programming" [Doets and van Eijck, 2004].

2.3.1 Project: DSLs, sets and von Neumann

This subsection describes a larger exercise (or small project) you can use to practice what you have learnt so far. In this project you will build up a domain-specific language (a DSL) for finite sets. The domain you should model is pure set theory where all members are sets.

Define a datatype *TERM* v for the abstract syntax of set expressions with variables of type v (as in Section 1.7.3) and a datatype *PRED* v for predicates over pure set expressions.

Part 1: *TERM* should have constructors for

- the *Empty* set

- the one-element set constructor *Singleton*

- *Union*, and *Intersection*

  - you can also try *Powerset*

- set-valued variables $(Var :: v \rightarrow TERM\ v)$

PRED should have constructors for

- the two predicates *Elem*, *Subset*

- the logical connectives *And*, *Or*, *Implies*, *Not*

Part 2:   A possible semantic domain for pure sets is

> **newtype** $Set = S\,[Set]$

Implement the evaluation functions

$$eval\ ::\ Eq\ v \Rightarrow Env\ v\ Set \to TERM\ v \to Set$$
$$check :: Eq\ v \Rightarrow Env\ v\ Set \to PRED\ v \to Bool$$

> **type** $Env\ var\ dom = [(var, dom)]$

Note that the type parameter v to $TERM$ is for the type of variables in the set expressions, not the type of elements of the sets. (You can think of pure set theory as "untyped" or "unityped".)

Part 3:   The *von Neumann encoding* of natural numbers as sets is defined recursively as

$$vonNeumann\ 0 \qquad\quad = Empty$$
$$vonNeumann\ (n+1) = Union\ (vonNeumann\ n)$$
$$\qquad\qquad\qquad\qquad (Singleton\ (vonNeumann\ n))$$

Implement *vonNeumann* and explore, explain and implement the following "pseudocode" claims as functions in Haskell:

$$claim1\ n1\ n2 = \{\text{- if } (n1 \leqslant n2) \text{ then } (n1 \subseteq n2) \text{ -}\}$$
$$claim2\ n \qquad = \{\text{- } n = \{0, 1, \ldots, n-1\} \text{ -}\}$$

You need to insert some embeddings and types and you should use the *eval* and *check* functions. (For debugging it is useful to implement a *show* function for *Set* which uses numerals to show the von Neumann naturals.)

2.4   Example proofs: contradiction, cases, primes

2.4.1   Proof by contradiction

Let us express and prove the irrationality of the square root of 2. We have two main concepts involved: the predicate "irrational" and the function "square root of". The square root function can be specified by the relation between two positive real numbers r and s as $r = \sqrt{s}$ iff $r\textasciicircum 2 == s$. The formula "x is irrational" is just $\neg (R\ x)$ where R is the predicate "is rational".[5]

$$R\ x = \exists\, a : \mathbb{Z}.\ \exists\, b : \mathbb{N}_{>0}.\ b * x == a\ \&\ GCD\ (a, b) == 1$$

The pattern "proof by contradiction" says that a way to prove some statement P is to assume $\neg\ P$ and derive something absurd. The traditional "absurd" statement is to prove simultaneously some Q and $\neg\ Q$, for example.

Let us take $P = \neg (R\ r)$ so that $\neg\ P = \neg\ (\neg\ (R\ r)) = R\ r$ and try with $Q = GCD\ (a, b) == 1$. Assuming $\neg\ P$ we immediately get Q so what we need is to prove $\neg\ Q$, that is $GCD\ (a, b) \neq 1$. We can use the equations $b * r == a$ and $r\textasciicircum 2 == 2$. Squaring the first equation and using the second we get $b\textasciicircum 2 * 2 == a\textasciicircum 2$. Thus $a\textasciicircum 2$ is even, which means that a is even, thus $a == 2 * c$ for some c. But then $b\textasciicircum 2 * 2 == a\textasciicircum 2 == 4 * c\textasciicircum 2$ which means that $b\textasciicircum 2 == 2 * c\textasciicircum 2$. By the same reasoning again we have that also b is even. But then 2 is a factor of both a and b, which means that $GCD\ (a, b) \geqslant 2$, which in turn implies $\neg\ Q$.

To sum up: by assuming $\neg\ P$ we can prove both Q and $\neg\ Q$. Thus, by contradiction, P must hold (the square root of two is irrational).

2.4.2 Proof by cases

As another example, let's prove that there are two irrational numbers p and q such that $p\textasciicircum q$ is rational.

$$S = \exists\, p.\ \exists\, q.\ \neg (R\ p)\ \&\ \neg (R\ q)\ \&\ R\ (p\textasciicircum q)$$

We know from above that $r = \sqrt{2}$ is irrational, so as a first attempt we could set $p = q = r$. Then we have satisfied two of the three clauses ($\neg (R\ p)$ and $\neg (R\ q)$). What about the third clause: is $x = p\textasciicircum q = r\textasciicircum r$ rational? By the principle of the excluded middle (Section 2.1.7), we know that either $R\ x$ or $\neg (R\ x)$ must hold. Then, we apply $\vee$-elimination, and thus we have to deal with the two possible cases separately.

Case 1: $R\ x$ holds. Then we have a proof of S with $p = q = r = \sqrt{2}$.

Case 2: $\neg (R\ x)$ holds. Then we have another irrational number x to play with. Let's try $p = x$ and $q = r$. Then $p\textasciicircum q = x\textasciicircum r = (r\textasciicircum r)\textasciicircum r = r\textasciicircum (r * r) = r\textasciicircum 2 = 2$ which is clearly rational. Thus, also in this case we have a proof of S, but now with $p = r\textasciicircum r$ and $q = r$.

---

[5]In fact we additionally require the rational to be normalised (no common factor between the denominator and numerator) to simplify the proof.

To sum up: yes, there are irrational numbers such that their power is rational. We can prove the existence without knowing what numbers p and q actually are: this is because negation-elimination is a *non-constructive principle*. The best we could do in an intuitionistic logic, which is constructive, is to show that, if they were not to exist, then we come to a contradiction.

2.4.3   There is always another prime

As an example of combining quantification (forall, exists) and implication let us turn to one statement of the fact that there are infinitely many primes. If we assume that we have a unary predicate expressing that a number is prime and a binary (infix) predicate ordering the natural numbers we can define a formula *IP* for "Infinitely many Primes" as follows:

$$IP = \forall\, n.\ Prime\ n \Rightarrow \exists\, m.\ Prime\ m\ \&\ m > n$$

Combined with the fact that there is at least one prime (like 2) we can repeatedly refer to this statement to produce a never-ending stream of primes.

To prove this formula we are going to translate from logic to programs as described in Section 2.1.6. We can translate step by step, starting from the top level. The forall-quantifier translates to a (dependent) function type $(n : Term) \rightarrow$ and the implication to a normal function type *Prime n* $\rightarrow$. The exists-quantifier translates to a (dependent) pair type $((m : Term), \ldots)$ and finally the & translates into a pair type. Putting all this together we get a type signature for any *proof* of the theorem:

$$proof : (n : Term) \rightarrow Prime\ n \rightarrow ((m : Term), (Prime\ m, m > n))$$

This time the proof is going to be constructive: we have to find a concrete bigger prime, m. We can start filling in the definition of *proof* as a 2-argument function returning a triple. The key idea is to consider $1 + factorial\ n$ as a candidate new prime:

$$proof\ n\ pn = (m, (pm, gt))$$
$$\mathbf{where}\ m' = 1 + factorial\ n$$
$$\qquad m = \{\text{- some non-trivial prime factor of } m' \text{ -}\}$$
$$\qquad pm = \{\text{- a proof that } m \text{ is prime -}\}$$
$$\qquad gt = \{\text{- a proof that } m > n \text{ -}\}$$

The proof *pm* is the core of the theorem. Let $x\%y$ be the remainder after integer division of x by y and let $(==_p)$ denote "equality modulo p": $x ==_p y = x\%p == y\%p$. Then note that for any number p where $2 \leqslant p \leqslant n$ we have $n! ==_p 0$. We then calculate

m' $==_p$ {- Def. of m' -}
$1 + n!$ $==_p$ {- modulo distributes over $+$, and $n!$ $==_p$ 0 -}
1

Thus m' is not divisible by any number from 2 to n. But is it a prime? Here we could, as before, use the law of excluded middle to progress. But we don't have to, because primality is a *decidable property*: we can write a terminating function which checks if m' is prime. We can then proceed by case analysis again: If m' is prime then $m = m'$ and the proof is done (because $1 + n! \geqslant 1 + n > n$). Otherwise, let m be a prime factor of m' (thus $m' = m * q, q > 1$). Then $1 ==_p m' ==_p (m\%p) * (q\%p)$ which means that neither m nor q are divisible by p (otherwise the product would be zero). As m is thus not divisible by any number from 2 to n, it has to be bigger than n. QED.

The constructive character of this proofs means that it can be used to define a (mildly useful) function which takes any prime number to some larger prime number. We can compute a few example values:

$2 \mapsto$ 3   (1+2!)
$3 \mapsto$ 7   (1+3!)
$5 \mapsto$ 11   (1+5! = 121 = 11*11)
$7 \mapsto$ 71   ...

2.5 Basic concepts of calculus

Now we have built up quite a bit of machinery to express logic formulas and proofs. It is time to apply it to some concepts in calculus. We start with the concept of "limit point" which is used in the formulation of different properties of limits of functions.

2.5.1 Limit point

The motivation comes from the expression $\lim_{x \to a} f(x)$ where $f : X \to \mathbb{R}$. When trying to formalise this, it turns out that not all combinations of X and a make sense. For example, with $f\, x = x\, /\, \sqrt{x}$ and $X = \mathbb{R}_{>0}$ it makes sense to ask for the limit at any $a : \mathbb{R}_{>0}$ and for $a = 0$, but not for $a = -1$, for example. The point a needs to be "approachable" from within X.

Definition 2.1. (adapted from Rudin [1964], page 28): Let X be a subset of $\mathbb{R}$. A point $p \in \mathbb{R}$ is a limit point of X iff for every $\epsilon > 0$, there exists $q \in X$ such that $q \neq p$ and $|q - p| < \epsilon$.

We will formalise this, starting with the types. To express "Let X be a subset of $\mathbb{R}$" we write $X : \mathcal{P}\,\mathbb{R}$. In general, the function $\mathcal{P}$ takes a set (here $\mathbb{R}$) to the set of all its subsets.

$Limp : \mathbb{R} \to \mathcal{P}\,\mathbb{R} \to Prop$
$Limp\ p\ X = \forall \epsilon > 0.\ \exists q \in X \smallsetminus \{p\}.\ |q - p| < \epsilon$

Notice that q depends on ϵ. Thus by introducing a function $getq$ we can move the $\exists$ out.

type $Q = \mathbb{R}_{>0} \to (X \smallsetminus \{p\})$
$Limp\ p\ X = \exists getq : Q.\ \forall \epsilon > 0.\ |getq\ \epsilon - p| < \epsilon$

Next, we introduce the function B such that $B\ c\ r$ is an "open ball" around c of radius r. On the real line this open ball is just an open interval, but with complex c or in more dimensions the term feels more natural.

$B : \mathbb{R} \to \mathbb{R}_{>0} \to \mathcal{P}\,\mathbb{R}$
$B\ c\ r = \{x \mid |x - c| < r\}$

Using B we get

$Limp\ p\ X = \exists getq : Q.\ \forall \epsilon > 0.\ getq\ \epsilon \in B\ p\ \epsilon$

Example Is $p = 1$ a limit point of $X = \{1\}$? No! $X \smallsetminus \{p\} = \{\}$ (there is no $q \neq p$ in X), thus there cannot exist a function $getq$ because it would have to return elements in the empty set!

Example Is $p = 1$ a limit point of the open interval $X = (0, 1)$? First note that $p \notin X$, but it is "very close" to X. A proof needs a function $getq$ which from any ϵ computes a point $q = getq\ \epsilon$ which is in both X and $B\ 1\ \epsilon$. We need a point q which is in X and *closer* than ϵ from 1. We can try with $q = 1 - \epsilon\,/\,2$ because $|1 - (1 - \epsilon\,/\,2)| = |\epsilon\,/\,2| = \epsilon\,/\,2 < \epsilon$ which means $q \in B\ 1\ \epsilon$. We also see that $q \neq 1$ because $\epsilon > 0$. The only remaining thing to check is that $q \in X$. This is true for sufficiently small ϵ but the function $getq$ must work for all positive reals. We can use any value in X (for example $17\,/\,38$) for ϵ which are "too big" ($\epsilon \geqslant 2$). Thus our function can be

$getq\ \epsilon \mid \epsilon < 2\quad\ = 1 - \epsilon\,/\,2$
$\quad\quad\quad\ \mid otherwise = 17\,/\,38$

A slight variation which is often useful would be to use *max* to define $getq\ \epsilon = max\ (17\,/\,38, 1 - \epsilon\,/\,2)$.

Similarly, we can show that any internal point (like $1\,/\,2$) is a limit point.

Example   Limit of an infinite discrete set $X = \{1 \, / \, n \mid n \in \mathbb{N}_{>0}\}$. Show that 0 is a limit point of X. Note (as above) that $0 \notin X$.

We want to prove *Limp* 0 X which is the same as $\exists getq : Q. \ \forall \epsilon > 0. \ getq \ \epsilon \in B \ 0 \ \epsilon$. Thus, we need a function *getq* which takes any $\epsilon > 0$ to an element of $X \setminus \{0\} = X$ which is less than ϵ away from 0. Or, equivalently, we need a function $getn : \mathbb{R}_{>0} \to \mathbb{N}_{>0}$ such that $1 \, / \, n < \epsilon$. Thus, we need to find an n such that $1 \, / \, \epsilon < n$. If $1 \, / \, \epsilon$ would be an integer we could use the next integer $(1 + 1 \, / \, \epsilon)$, so the only step remaining is to round up:

$$getq \ \epsilon = 1 \, / \, getn \ \epsilon$$
$$getn \ \epsilon = 1 + ceiling \ (1 \, / \, \epsilon)$$

Exercise 2.7. prove that 0 is the *only* limit point of X.

Proposition 2.1. If X is finite, then it has no limit points:

$$\forall p \in \mathbb{R}. \ \neg \ (Limp \ p \ X)$$

Proof. This is a good exercise in quantifier negation!

$$
\begin{aligned}
&\neg \ (Limp \ p \ X) & &= \{\text{- Def. of } Limp \text{ -}\} \\
&\neg \ (\exists getq : Q. \ \forall \epsilon > 0. \ getq \ \epsilon \in B \ p \ \epsilon) & &= \{\text{- Negation of existential -}\} \\
&\forall getq : Q. \ \neg \ (\forall \epsilon > 0. \ getq \ \epsilon \in B \ p \ \epsilon) & &= \{\text{- Negation of universal -}\} \\
&\forall getq : Q. \ \exists \epsilon > 0. \ \neg \ (getq \ \epsilon \in B \ p \ \epsilon) & &= \{\text{- Simplification -}\} \\
&\forall getq : Q. \ \exists \epsilon > 0. \ |getq \ \epsilon - p| \geqslant \epsilon
\end{aligned}
$$

Thus, using the "functional interpretation" of this type we see that a proof needs a function *noLim*

$$noLim : (getq : Q) \to \mathbb{R}_{>0}$$

such that **let** $\epsilon = noLim \ getq$ **in** $|getq \ \epsilon - p| \geqslant \epsilon$.

Note that *noLim* is a *higher-order* function: it takes a function *getq* as an argument. How can we analyse this function to find a suitable ϵ? The key here is that the range of *getq* is $X \setminus \{p\}$ which is a finite set (not containing p). Thus we can enumerate all the possible results in a list $xs = [x_1, x_2, \ldots x_n]$, and measure their distances to p: $ds = map \ (\lambda x \to |x - p|) \ xs$. Now, if we let $\epsilon = minimum \ ds$ we can be certain that $|getq \ \epsilon - p| \geqslant \epsilon$ just as required (and $\epsilon \neq 0$ because $p \notin xs$). □

Exercise 2.8. Show that *Limp* p X implies that X is infinite.

Show how to construct an infinite sequence $a : \mathbb{N} \to \mathbb{R}$ of points in $X \setminus \{p\}$ which gets arbitrarily close to p. Note that this construction can be seen as a proof of *Limp* p $X \Rightarrow$ *Infinite* X.

2.5.2   The limit of a sequence

Now we can move from limit points to the more familiar limit of a sequence. At the core of this book is the ability to analyse definitions from mathematical texts, and here we will use the definition of the limit of a sequence of Adams and Essex [2010, page 498]:

> We say that sequence a_n converges to the limit L, and we write $\lim_{n\to\infty} a_n = L$, if for every positive real number ϵ there exists an integer N (which may depend on ϵ) such that if $n > N$, then $|a_n - L| < \epsilon$.

The first step is to type the variables introduced. A sequence a is a function from $\mathbb{N}$ to $\mathbb{R}$, thus $a : \mathbb{N} \to \mathbb{R}$ where a_n is special syntax for normal function application of a to $n : \mathbb{N}$. Then we have $L : \mathbb{R}$, $\epsilon : \mathbb{R}_{>0}$, and $N : \mathbb{N}$ (or $getN : \mathbb{R}_{>0} \to \mathbb{N}$ as we will see later).

In the next step we analyse the new concept introduced: the syntactic form $\lim_{n\to\infty} a_n = L$ which we could express as an infix binary predicate *haslim* where *a haslim L* is well-typed if $a : \mathbb{N} \to \mathbb{R}$ and $L : \mathbb{R}$. Note that the equality sign is abused in the traditional mathematical notation: it looks like *lim* would be a normal function always returning a $\mathbb{R}$, where in fact it is not always defined. As mentioned in Section 1.3, one way to handle is this to treat *lim* as a partial function, modelled in Haskell as returning *Maybe* $\mathbb{R}$. Here we play it safe and use a relation instead — because at this stage we cannot be sure if the limit is even unique.

The third step is to formalise the definition using logic: we define *haslim* using a ternary helper predicate P:

$$a \ haslim \ L = \forall \epsilon > 0.\ P \ a \ L \ \epsilon \quad \text{-- ``for every positive real number } \epsilon \ldots\text{''}$$
$$\begin{aligned}P \ a \ L \ \epsilon &= \exists N : \mathbb{N}.\ \forall n \geq N.\ |a_n - L| < \epsilon \\ &= \exists N : \mathbb{N}.\ \forall n \geq N.\ a_n \in B \ L \ \epsilon \\ &= \exists N : \mathbb{N}.\ I \ a \ N \subseteq B \ L \ \epsilon\end{aligned}$$

where we have introduced an "image function" for sequences "from N onward":

$$I : (\mathbb{N} \to X) \to \mathbb{N} \to \mathcal{P} \ X$$
$$I \ a \ N = \{a \ n \mid n \geq N\}$$

The "forall-exists"-pattern is very common and it is often useful to transform such formulas into another form. In general $\forall x : X.\ \exists y : Y.\ Q \ x \ y$ is equivalent to $\exists\, gety : X \to Y.\ \forall\, x : X.\ Q\, x\ (gety\ x)$. In the new form we more clearly see the function *gety* which shows how the choice of y depends on x. For our case with *haslim* we can thus write

$$a\ haslim\ L\ =\ \exists\, getN : \mathbb{R}_{>0} \to \mathbb{N}.\ \forall\, \epsilon > 0.\ I\, a\ (getN\ \epsilon) \subseteq B\, L\, \epsilon$$

where we have made the function *getN* more visible. The core evidence of *a haslim L* is the existence of such a function (with suitable properties).

Exercise 2.9. Prove that the limit of a sequence is unique.

Exercise 2.10. Prove $(a\ haslim\ L)\ \&\ (b\ haslim\ M) \Rightarrow (a+b)\ haslim\ (L+M)$.

When we are not interested in the exact limit, just that it exists, we say that a sequence a is *convergent* when $\exists L.\ a\ haslim\ L$.

2.5.3 Case study: The limit of a function

As our next mathematical text book quote we take the definition of the limit of a function of type $\mathbb{R} \to \mathbb{R}$ from Adams and Essex [2010]:

A formal definition of limit

We say that $f(x)$ **approaches the limit** L as x **approaches** a, and we write

$$\lim_{x \to a} f(x) = L,$$

if the following condition is satisfied:
for every number $\epsilon > 0$ there exists a number $\delta > 0$, possibly depending on ϵ, such that if $0 < |x - a| < \delta$, then x belongs to the domain of f and

$$|f(x) - L| < \epsilon.$$

The *lim* notation has four components: a variable name x, a point a, an expression $f(x)$ and the limit L. The variable name and the expression can be combined into just the function f[6] and this leaves us with three essential components: $a, f,$ and L. Thus, *lim* can be seen as a ternary (3-argument) predicate which is satisfied if the limit of f exists at a and equals L. If we apply our logic toolbox we can define *lim* starting something like this:

$$lim\ a\ f\ L = \forall\, \epsilon > 0.\ \exists\, \delta > 0.\ P\, \epsilon\, \delta$$

---

[6]To see why this works in the general case of any expression of x, read Section 1.7

when P is a predicate yet to define. Indeed, it is often useful to introduce a local name (like P here) to help break the definition down into more manageable parts. If we now naively translate the last part we get this "definition" for P:

$$\textbf{where } P \; \epsilon \; \delta = (0 < |x - a| < \delta) \Rightarrow (x \in Dom \, f \land |f \, x - L| < \epsilon))$$

Note that there is a scoping problem: we have a, f, and L from the "call" to lim and we have ϵ and δ from the two quantifiers, but where did x come from? It turns out that the formulation "if … then …" hides a quantifier that binds x. Thus we get this definition:

$$lim \; a \; f \; L = \forall \, \epsilon > 0. \; \exists \, \delta > 0. \; \forall \, x. \; P \, \epsilon \, \delta \, x$$
$$\textbf{where } P \; \epsilon \; \delta \; x = (0 < |x - a| < \delta) \Rightarrow (x \in Dom \, f \land |f \, x - L| < \epsilon))$$

The predicate lim can be shown to be a partial function of two arguments, a and f. This means that at a point a each function f can have *at most* one limit L. (This is not evident from the definition and proving it is a good exercise.)

Exercise 2.11. What does Adams mean by "$\delta > 0$, possibly depending on ϵ"? How did we express "possibly depending on" in our formal definition? Hint: how would you express that δ cannot depend on ϵ?

2.6 Exercises

2.6.1 Representations of propositions

Exercise 2.12. Define a function for De Morgan dualisation.

Exercise 2.13. Define a function to rewrite propositions into conjunctive normal form.

Exercise 2.14. Define a function to rewrite propositions into disjunctive normal form.

Exercise 2.15. Propositions as polynomials. (This is a difficult exercise: it is a good idea to come back to it after Chapter 4 (where one learns about abstract structures) and after Chapter 5).

One way to connect logic to calculus is to view propositions as polynomials (in several variables). The key idea is to represent the truth values by zero (False) and one (True) and each named proposition P by a fresh variable p.

To represent logical operations one just has to check that the usual notion of expression evaluation gives the right answer for zero and one. (Can you express this as a homomorphism — seen in Chapter 4?)

The simplest operation to represent is *And* which becomes multiplication: the predicate *And P Q* translates to $p * q$ as can be easily checked. Note that $p + q$ does not represent any proposition, because its value would be 2 for $p = q = 1$, but 2 does not represent any Boolean.

How should *Not*, *Or*, and *Implies* be represented?

2.6.2 Proofs

```
{-# LANGUAGE EmptyCase #-}
import PropositionalLogic
```

Short technical note   For the exercises on the abstract representation of proofs for the propositional calculus using Haskell, (see Section 2.1.5), you might find it useful to take a look at typed holes, a feature which is enabled by default in GHC and available (the same way as the language extension EmptyCase above) from version 7.8.1 onwards: https://wiki.haskell.org/GHC/Typed_holes.

If you are familiar with Agda, these will be familiar to use. In summary, when trying to code up the definition of some expression (which you have already typed) you can get GHC's type-checker to help you out a little in seeing how far you might be from forming the expression you want. That is, how far you are from constructing something of the appropriate type.

Take *example0* below, and say you are writing:

> *example0 e = andIntro (_ e) _*

When loading the module, GHC will tell you which types your holes (marked by "_") should have for the expression to be type correct.

On to the exercises.

Exercise 2.16. Prove these theorems (for arbitrary p, q and r):

> *Impl (And p q) q*
> *Or p q → Or q p*
> *(p → q) → (Not q → Not p)*   -- call it *notMap*
> *Or p (Not p)*   -- recall the law of excluded middle

For the hardest examples it can be good to use "theory exploration": try to combine the earlier theorems and rules to build up suitable term for which *notMap* or *notElim* could be used.

Exercise 2.17. Translate to Haskell and prove the De Morgan laws:

$$\neg (p \lor q) \leftrightarrow \neg p \land \neg q$$
$$\neg (p \land q) \leftrightarrow \neg p \lor \neg q$$

(translate equivalence to conjunction of two implications).

Exercise 2.18. So far, the implementation of the datatypes has played no role: we treated them as abstract. To make this clearer: define the types for connectives in *AbstractFol* in any way you wish, e.g.:

> **newtype** *And p q = A p q*
> **newtype** *Not p   = B p*

etc. as long as you still export only the data types, and not the constructors. Convince yourself that the proofs given above still work and that the type-checker can indeed be used as a poor man's proof-checker.

Exercise 2.19. From now on you can assume the representation of proofs defined in Section 2.1.6.

1. Check your understanding by redefining all the introduction and elimination rules as functions.

2. Compare proving the distributivity laws

$$(p \wedge q) \vee r \leftrightarrow (p \vee r) \wedge (q \vee r)$$
$$(p \vee q) \wedge r \leftrightarrow (p \wedge r) \vee (q \wedge r)$$

using only the introduction and elimination rules (no pairs, functions, etc.), with writing the corresponding functions with the given implementations of the datatypes. The first law, for example, requires a pair of functions:

$$(Either\ (p,q)\ r \rightarrow (Either\ p\ r, Either\ q\ r)$$
$$,(Either\ p\ r, Either\ q\ r) \rightarrow Either\ (p,q)\ r$$
$$)$$

Exercise 2.20. Assume

type $Not\ p = p \rightarrow False$

Implement *notIntro2* using the definition of *Not* above, i.e., find a function

$notIntro2 :: (p \rightarrow (q, q \rightarrow False)) \rightarrow (p \rightarrow False)$

Using

$contraHey :: False \rightarrow p$
$contraHey\ evE = \textbf{case}\ evE\ \textbf{of}\ \{\ \}$

prove

$$(q \wedge \neg q) \rightarrow p$$

Can you prove $p \vee \neg p$?

Prove

$$\neg p \vee \neg q \rightarrow \neg(p \wedge q)$$

Can you prove the converse?

Exercise 2.21. Recall that every sentence provable in constructive logic is provable in classical logic. But the converse, as we have seen in the previous exercise, does not hold. On the other hand, there is no sentence in classical logic which would be contradicted in constructive logic. In particular, while we cannot prove $p \vee \neg p$, we *can* prove (constructively!) that there is no p for which $\neg(p \vee \neg p)$, i.e., that the sentence $\neg \neg (p \vee \neg p)$ is always true.

Show this by implementing the following function:

$$noContra :: (Either\ p\ (p \to False) \to False) \to False$$

Hint: The key is to use the function argument to *noContra* twice.

2.6.3   Continuity and limits

Below, when asked to "sketch an implementation" of a function, you must explain how the various results might be obtained from the arguments, in particular, why the evidence required as output may result from the evidence given as input. You may use all the facts you know (for instance, that addition is monotonic) without formalisation.

Exercise 2.22. Consider the classical definition of continuity:

> *Definition:* Let $X \subseteq \mathbb{R}$, and $c \in X$. A function $f : X \to \mathbb{R}$ is *continuous at* c if for every $\epsilon > 0$, there exists $\delta > 0$ such that, for every x in the domain of f, if $|x - c| < \delta$, then $|fx - fc| < \epsilon$.

1. Write the definition formally, using logical connectives and quantifiers.

2. Introduce functions and types to simplify the definition.

3. Prove the following proposition: If f and g are continuous at c, f + g is continuous at c.

Exercise 2.23. Adequate notation for mathematical concepts and proofs.

A formal definition of "$f : X \to \mathbb{R}$ is continuous" and "f is continuous at c" can be written as follows (using the helper predicate Q):

$$
\begin{aligned}
C\ (f) &= \forall\, c : X.\ Cat\ (f, c) \\
Cat\ (f, c) &= \forall\, \epsilon > 0.\ \exists\, \delta > 0.\ Q\ (f, c, \epsilon, \delta) \\
Q\ (f, c, \epsilon, \delta) &= \forall\, x : X.\ |x - c| < \delta \Rightarrow |f\,x - f\,c| < \epsilon
\end{aligned}
$$

By moving the existential quantifier outwards we can introduce the function $get\delta$ which computes the required δ from c and ϵ:

$$C'\ (f) = \exists\, get\delta : X \to \mathbb{R}_{>0} \to \mathbb{R}_{>0}.\ \forall c : X.\ \forall \epsilon > 0.\ Q\ (f, c, \epsilon, get\delta\ c\ \epsilon)$$

Now, consider this definition of *uniform continuity*:

> **Definition:** Let $X \subseteq \mathbb{R}$. A function $f : X \to \mathbb{R}$ is *uniformly continuous* if for every $\epsilon > 0$, there exists $\delta > 0$ such that, for every x and y in the domain of f, if $|x - y| < \delta$, then $|f\ x - f\ y| < \epsilon$.

1. Write the definition of $UC\ (f) =$ "f is uniformly continuous" formally, using logical connectives and quantifiers. Try to use Q.

2. Transform $UC\ (f)$ into a new definition $UC'\ (f)$ by a transformation similar to the one from $C\ (f)$ to $C'\ (f)$. Explain the new function $new\delta$ introduced.

3. Prove that $\forall f : X \to \mathbb{R}.\ UC'\ (f) \Rightarrow C'\ (f)$. Explain your reasoning in terms of $get\delta$ and $new\delta$.

Exercise 2.24. Consider the statement:

The sequence $\{a_n\} = (0, 1, 0, 1, \ldots)$ does not converge.

To keep things short, let us abbreviate a significant chunk of the definition of *a haslim L* (see Section 2.5.2) by

$$P : Seq\ X \to X \to \mathbb{R}_{>0} \to Prop$$
$$P\ a\ L\ \epsilon = \exists\, N : \mathbb{N}.\ \forall n : \mathbb{N}.\ (n \geq N) \to (|a_n - L| < \epsilon)$$

1. Define the sequence $\{a_n\}$ as a function $a : \mathbb{N} \to \mathbb{R}$.

2. The statement "the sequence $\{a_n\}$ is convergent" is formalised as

$$\exists\, L : \mathbb{R}.\ \forall \epsilon > 0.\ P\ a\ L\ \epsilon$$

   The formalisation of "the sequence $\{a_n\}$ is not convergent" is therefore

$$\neg\, \exists\, L : \mathbb{R}.\ \forall \epsilon > 0.\ P\ a\ L\ \epsilon$$

   Simplify this expression using the rules

$$\neg\, (\exists x.\ P\ x) \leftrightarrow (\forall x.\ \neg\, (P\ x))$$
$$\neg\, (\forall x.\ P\ x) \leftrightarrow (\exists x.\ \neg\, (P\ x))$$
$$\neg\, (P \to Q) \leftrightarrow P \wedge \neg Q$$

   The resulting formula should have no $\neg$ in it (that's possible because the negation of $<$ is $\geq$).

3. Give a functional interpretation of the resulting formula.

4. Sketch an implementation of the function, considering two cases: $L \neq 0$ and $L = 0$.

Exercise 2.25. Same as Exercise 2.24 but for $a = id$.

Exercise 2.26. Consider the statement:

The limit of a convergent sequence is unique.

1. There are many ways of formalising this in FOL. For example:

   let $Q \, a \, L = \forall \, \epsilon > 0. \; P \, a \, L \, \epsilon$
   in $\forall \, L_1 : \mathbb{R}. \; \forall \, L_2 : \mathbb{R}. \; (Q \, a \, L_1 \wedge Q \, a \, L_2) \rightarrow L_1 = L_2$

   i.e., if the sequence converges to two limits, then they must be equal, or

   $\forall \, L_1 : \mathbb{R}. \; \forall \, L_2 : \mathbb{R}. \; Q \, a \, L_1 \wedge L_1 \neq L_2 \rightarrow \neg \, Q \, a \, L_2$

   i.e., if a sequence converges to a limit, then it doesn't converge to anything that isn't the limit.

   Simplify the latter alternative to eliminate the negation and give functional representations of both.

2. Choose one of the functions and sketch an implementation of it.

Chapter 3

Types in Mathematics

In this chapter we continue the quest to assign types to mathematical concepts. In Section 3.1 through Section 3.4 we go through several examples of short snippets from mathematical texts, with different kinds of derivatives at the centre. We also show (in Section 3.5) how to collect related types in Haskell's type classes and define several classes for the "numerical hierarchy": *Additive*, *AddGroup*, *Ring*, *Field*, etc. These classes provide generalised versions of the standard arithmetic operations like $(+)$, $(-)$, $(*)$, $(/)$, etc. in preparation for the coming chapters. Finally, in Section 3.6, we use equational reasoning to calculate a Haskell version of the classical derivative rules (for sums, products, etc.).

3.1 Typing Mathematics: derivative of a function

Consider the classical definition of the derivative of Adams and Essex [2010]:

> The **derivative** of a function f is another function f' defined by
>
> $$f'(x) = \lim_{h \to 0} \frac{f(x+h) - f(x)}{h}$$
>
> at all points x for which the limit exists (i.e., is a finite real number). If $f'(x)$ exists, we say that f is **differentiable** at x.

We can start by assigning types to the expressions in the definition. Let's write X for the domain of f so that we have $f : X \to \mathbb{R}$ and $X \subseteq \mathbb{R}$ (or, equivalently,

81

$X : \mathcal{P} \, \mathbb{R}$). If we denote with Y the subset of X for which f is differentiable we get $f' : Y \to \mathbb{R}$. Thus, the operation which maps f to f' has type $(X \to \mathbb{R}) \to (Y \to \mathbb{R})$. Unfortunately, the only notation for this operation given (implicitly) in the definition is a prime symbol (apostrophe), written postfix. To make it easier to see we use a prefix D instead and we can thus write $D : (X \to \mathbb{R}) \to (Y \to \mathbb{R})$. We will often assume that $X = Y$ (f is differentiable everywhere) so that we can see D as preserving the type of its argument.

Now, with the type of D sorted out, we can turn to the actual definition of the function $D\,f$. The definition is given for a fixed (but arbitrary) x. (At this point the reader may want to check the definition of "limit of a function" in Section 2.5.3.) The *lim* expression is using the (anonymous) function $g\ h = \frac{f(x+h)-f\,x}{h}$ and that the limit of g is taken at 0. Note that g is defined in the scope of x and that its definition uses x so it can be seen as having x as an implicit, first argument. To be more explicit we write $\varphi\ x\ h = \frac{f(x+h)-f\,x}{h}$ and take the limit of $\varphi\ x$ at 0. So, to sum up, $D\,f\,x = lim\ 0\ (\varphi\ x)$.[1] The key here is that we name, type, and specify the operation of computing the derivative (of a one-argument function). We will use this operation quite a bit in the rest of the book, but here are just a few examples to get used to the notation. With the following definitions:

$$
\begin{aligned}
sq\ x &= x\textasciicircum 2 \\
double\ x &= 2 * x \\
c_2\ x &= 2
\end{aligned}
$$

we have the following equalities:

$$
\begin{aligned}
sq' &\mathrel{==} D\ sq \mathrel{==} D\ (\lambda x \to x\textasciicircum 2) \mathrel{==} D\ (\textasciicircum 2) \mathrel{==} (2*) \mathrel{==} double \\
sq'' &\mathrel{==} D\ sq' \mathrel{==} D\ double \mathrel{==} c_2 \mathrel{==} const\ 2
\end{aligned}
$$

What we cannot do at this stage is to actually *implement* D in Haskell. If we only have a function $f : \mathbb{R} \to \mathbb{R}$ as a "black box" we cannot really compute the actual derivative $f' : \mathbb{R} \to \mathbb{R}$, but only numerical approximations. However if we also have access to the "source code" of f, then we can apply the usual rules we have learnt in calculus. We will get back to this question in Section 3.6.

3.2 Typing Mathematics: partial derivative

Armed with our knowledge of functions of more than one variable, we can continue on our quest to type the elements of mathematical textbook defini-

---

[1] We could go one step further by noting that f is in the scope of φ and used in its definition. Thus the function $\psi\,f\,x\,h = \varphi\,x\,h$, or $\psi\,f = \varphi$, is used. With this notation we obtain a point-free definition that can come in handy: $D\,f = lim\ 0 \circ \psi\,f$.

tions. Our example here is by Mac Lane [1986, page 169], where we read

1   [...] a function $z = f(x, y)$ for all points (x, y) in some open set U of
2   the Cartesian (x, y)-plane. [...] If one holds y fixed, the quantity z
3   remains just a function of x; its derivative, when it exists, is called
4   the *partial derivative* with respect to x. Thus at a point (x, y) in U
5   this derivative for $h \neq 0$ is

$$\partial z / \partial x = f'_x(x, y) = \lim_{h \to 0} (f(x + h, y) - f(x, y))/h$$

What are the types of the elements involved? We have

$$U \subseteq \mathbb{R} \times \mathbb{R} \quad \text{-- Cartesian plane}$$
$$f \; : U \to \mathbb{R}$$
$$z \; : U \to \mathbb{R} \quad \text{-- but see below}$$
$$f'_x : U \to \mathbb{R}$$

The x in the subscript of f' is *not* a real number, but a symbol (we used *String* for similar purposes in Section 1.7.3).

The expression (x, y) has six occurrences. The first two (on line 1) denote variables of type U, the third (on line 2) is just a name ((x, y)-plane). The fourth (at line 4) denotes a variable of type U bound by a universal quantifier: "a point (x, y) in U" as text which would translate to $\forall (x, y) \in U$ as a formula fragment.

The variable h is a non-zero real number. The use of the word "for" might lead one to believe that it is bound by a universal quantifier ("for $h \neq 0$" on line 4), but that is incorrect. In fact, h is used as a local variable introduced in the subscript of lim. This variable h is a parameter of an anonymous function, whose limit is then taken at 0.

That function, which we can name φ, has the type $\varphi : U \to (\mathbb{R} \setminus \{0\}) \to \mathbb{R}$ and is defined by

$$\varphi \, (x, y) \, h = (f \, (x + h, y) - f \, (x, y)) \, / \, h$$

The limit is then written *lim* $0 \, (\varphi \, (x, y))$. Note that 0 is a limit point of $\mathbb{R} \setminus \{0\}$, so the type of *lim* is the one we have discussed:

$$lim : \{p \mid p \in \mathbb{R}, Limp \; p \; X\} \to (X \to \mathbb{R}) \to \mathbb{R}$$

On line 1, $z = f \, (x, y)$ probably does not mean that we let z be a fixed value in $\mathbb{R}$, although the phrase "the quantity z" (on line 2) suggests this. Rather, a possible interpretation is that z is used to abbreviate the expression f (x, y). That is, z stands for an expression which depends on x and y; thus, it can be enlightening to replace z with f (x, y) everywhere. In particular, $\partial z / \partial x$ becomes ∂f $(x, y) / \partial x$, which we can interpret as the operator $\partial / \partial x$ applied to f and (x, y) (remember that (x, y) is bound in the context by a universal quantifier on line 4). There is the added difficulty that, just like the subscript in f'_x, the x in ∂x is not the x bound by the universal quantifier, but just a symbol.

To sum up, partial derivative operators which mention symbols (such as $\partial / \partial x$ or "prime subscript x") do act on an representation of functions which uses symbols for the variables (not positions), such as presented in Section 1.7.3. This is why we mostly see $\partial f / \partial x$, $\partial f / \partial y$, $\partial f / \partial z$ etc. when, in the context, the function f has been given a definition of the form $f(x, y, z) = \ldots$. This kind of approach presents several difficulties:

1. it makes it hard to rename variables (for example for the purpose of integration)

2. Further confusion can be created when a variable (such as z above) depends on other variables. Tracing dependencies can become daunting and it is easy to make errors of name when doing calculations.

3. it makes it difficult to assign a higher-order type to the partial derivatives. Indeed, as we have seen in Section 1.7.2, the $\partial f / \partial x$ style means that the operator binds the name of the variable.

One possibility would be to use the following type: $\partial / \partial x_i : (\mathbb{R}^n \to \mathbb{R}) \to (\mathbb{R}^n \to \mathbb{R})$, but it still assumes as input a vector of variables x— even though the type assumes independence with respect to the variable names. Hence we prefer a notation which doesn't rely on the names given to the arguments whatsoever. It was popularised by Landau [1934] (English edition Landau [2001]): D_1 for the partial derivative with respect to the first argument, D_2 for the partial derivative with respect to the second argument, etc.

Exercise 3.1. Partial Derivatives For $f : \mathbb{R}^2 \to \mathbb{R}$ define D_1 and D_2 using only D. In more detail: let the type $F2 = \mathbb{R}^2 \to \mathbb{R}$ and $F1 = \mathbb{R} \to \mathbb{R}$. Then both D_1 and D_2 have type $F2 \to F2$ and $D : F1 \to F1$. Start by defining helper functions:

$$fstFixed \ :: a \to (b \to (a, b))$$
$$sndFixed :: b \to (a \to (a, b))$$

Hint: there is only one type-correct definition of each. Then use D and the helpers in the definitions of D_1 and D_2.

3.3 Typing Mathematics: Lagrangian case study

From Sussman and Wisdom [2013]:

> A mechanical system is described by a Lagrangian function of the
> system state (time, coordinates, and velocities). A motion of the
> system is described by a path that gives the coordinates for each
> moment of time. A path is allowed if and only if it satisfies the La-
> grange equations. Traditionally, the Lagrange equations are writ-
> ten
>
> $$\frac{d}{dt}\frac{\partial L}{\partial \dot{q}} - \frac{\partial L}{\partial q} = 0$$
>
> What could this expression possibly mean?

To start answering the question of Sussman and Wisdom, we start typing the
elements involved:

1. First, note that the "system state" mentioned can be modelled as a triple
   (of type $S = T \times Q \times V$) and we can call the three components $t : T$ for
   time, $q : Q$ for coordinates, and $v : V$ for velocities.

2. If we let "coordinates" be just one coordinate, then there is also a single
   velocity. (A bit of physics domain knowledge is useful here: if q is a
   position of a particle, then v is its velocity.) Thus we can use $T = Q =
   V = \mathbb{R}$ in this example but it can help the reading to remember the
   different uses of $\mathbb{R}$ — this would help for example to generalise to more
   than one coordinate.

3. Also the use of notation for "partial derivative", $\partial L / \partial q$, suggests that L
   is a function of at least a pair of arguments:

   $L : \mathbb{R}^i \to \mathbb{R}, i \geq 2$

   This is consistent with our plan so far if we take $i = 3$:

   $L : \mathbb{R}^3 \to \mathbb{R}$

4. Looking again at the same derivative, $\partial L / \partial q$ suggests that q is the name
   of a real variable, one of the three arguments to L. In the context, which
   we do not have, we would expect to find somewhere the definition of
   the Lagrangian as a function of the system state:

   $L : T \times Q \times V \to \mathbb{R}$
   $L (t, q, v) = \ldots$

5. Consequently the type of the partial derivatives get specialised as follows:

$$(\partial/\partial q) : (T \times Q \times V \to \mathbb{R}) \to (T \times Q \times V \to \mathbb{R})$$

The notation $\partial L/\partial q$ is equivalent to $(\partial/\partial q)\ L$, and $D_2\ L$; applying the partial derivative with respect to the second argument (named q) of L.

6. Therefore, $\partial L/\partial q$ should also be a function of the same triple of arguments as L:

$$(\partial L/\partial q) : T \times Q \times V \to \mathbb{R}$$

It follows that the equation expresses a relation between *functions*, thus the 0 on the right-hand side of the Lagrange equation(s) is *not* the real number 0, but rather the constant function *const* 0:

$$const\ 0 : T \times Q \times V \to \mathbb{R}$$
$$const\ 0\quad (t, q, v)\quad = 0$$

7. We now have a problem: d/dt can only be applied to functions of *one* real argument t, and the result is a function of one real argument:

$$(d/dt)\ (\partial L/\partial \dot{q}) : T \to \mathbb{R}$$

Since we subtract from this the function $\partial L/\partial q$, it follows that this, too, must be of type $T \to \mathbb{R}$. But we already typed it as $T \times Q \times V \to \mathbb{R}$, contradiction!

8. The expression $\partial L/\partial \dot{q}$ appears to also be malformed. We would expect a variable name where we find $\dot{q}$, but $\dot{q}$ is the same as dq/dt, a function. But, with some knowledge from physics we can guess that $\dot{q}$, the rate of change of the position with time, is the same as the v, the velocity. Thus $\partial L/\partial \dot{q} = \partial L/\partial v = D_3\ L$ — now well-formed, but still ill-typed.

9. Looking back at the description above, we see that the only immediate candidate for an application of d/dt is "a path that gives the coordinates for each moment of time". Thus, the path is a function of time, let us say

$$w : T \to Q \quad \text{-- with } T \text{ for time and } Q \text{ for coordinates } (q : Q)$$

We can now guess that the use of the plural form "equations" might have something to do with the use of "coordinates" in the plural. In an n-dimensional space, a position is given by n coordinates. A path would then be a function

$$w : T \to Q \quad \text{-- with } Q = \mathbb{R}^n$$

which is equivalent to n functions of type $T \to \mathbb{R}$, each computing one coordinate as a function of time. We would then have an equation for each of them. But we will come back to use $n = 1$ for the rest of this example.

10. Now that we have a path, the coordinates at any time are given by the path. And because the time derivative of a coordinate is a velocity, we can actually compute the trajectory of the full system state (T, Q, V) starting from just the path.

$$q : T \to Q$$
$$q\, t = w\, t \qquad \text{-- or, equivalently, } q = w$$
$$\dot{q} : T \to V$$
$$\dot{q}\, t = dw/dt \quad \text{-- or, equivalently, } \dot{q} = D\, w$$

We combine these in the "combinator" *expand*, given by

$$expand : (T \to Q) \to (T \to T \times Q \times V)$$
$$expand\, w\, t = (t, w\, t, D\, w\, t)$$

11. With *expand* in our toolbox we can fix the typing problem in item 7 above. The Lagrangian is a "function of the system state (time, coordinates, and velocities)" and the "expanded path" (*expand w*) computes the state from just the time. By composing them we get a function

$$L \circ (expand\, w) : T \to \mathbb{R}$$

which describes how the Lagrangian would vary over time if the system would evolve according to the path w.

This particular composition is not used in the equation, but we do have

$$(\partial L / \partial q) \circ (expand\, w) : T \to \mathbb{R}$$

which is used inside d/dt (and which now type-checks).

12. We now move to using D for d/dt, D_2 for $\partial/\partial q$, and D_3 for $\partial/\partial \dot{q}$. The type of the partial derivatives D_2 and D_3 is $(S \to \mathbb{R}) \to (S \to \mathbb{R})$, and here $D : (T \to \mathbb{R}) \to (T \to \mathbb{R})$. In combination with *expand w* we find these type correct combinations for the two terms in the equation:

$$D\,((D_2\, L) \circ (expand\, w)) : T \to \mathbb{R}$$
$$(D_3\, L) \circ (expand\, w) \quad : T \to \mathbb{R}$$

The equation becomes

$$D\left((D_3\ L) \circ (expand\ w)\right) - (D_2\ L) \circ (expand\ w) = const\ 0$$

or, after simplification:

$$D\ (D_3\ L \circ expand\ w) = D_2\ L \circ expand\ w$$

where both sides are functions of type $T \to \mathbb{R}$.

13. "A path is allowed if and only if it satisfies the Lagrange equations" means that this equation is a predicate on paths (for a particular L):

$$Lagrange\ (L, w) = D\ (D_3\ L \circ expand\ w) \mathbin{==} D_2\ L \circ expand\ w$$

where we use ($==$) in the equation to avoid confusion with the equality sign ($=$) used for the definition of the predicate.

So, we have figured out what the equation means in terms of operators that we recognise. If we zoom out slightly we see that the quoted text means something like: If we can describe the mechanical system in terms of a "Lagrangian" ($L : S \to \mathbb{R}$ where $S = T \times Q \times V$), then we can use the equation to check if a particular candidate path $w : T \to \mathbb{R}$ qualifies as an allowed "motion of the system" or not. The unknown of the equation is the path w, and as the equation involves partial derivatives it is an example of a partial differential equation (a PDE). We will not dig into how to solve such PDEs, but they are widely used in physics.

In Exercise 3.6 you can practice checking under which conditions the Lagrange equations are satisfied for some candidates paths in a simple case of an object moving in constant gravity. If you have the right background in physics it is really instructive to try out a few more advanced examples of Lagrangians, or check the similar method of Hamiltonian dynamics described in Exercise 3.11 (which includes links to blog posts and Haskell libraries helping with simulations of the modelled systems).

3.4   Incremental analysis with types

So far in this chapter we have worked on typing mathematics, but without the help of any tool. However we can in fact get the Haskell interpreter to help a bit even when we are still at the specification stage — before we have any code running. It is often useful to collect the known (or assumed) facts about types in a Haskell file and regularly check if the type-checker agrees. This is a form of type-driven development and can help avoiding wrong turns even for concepts which cannot be fully implemented.

Consider the following quote from Mac Lane [1986, page 182]:

1. In these cases one tries to find not the values of x which make a
2. given function $y = f(x)$ a minimum, but the values of a given
3. function $f(x)$ which make a given quantity a minimum. Typically,
4. that quantity is usually measured by an integral whose integrand
5. is some expression F involving both x, values of the function $y = f(x)$ at interest and the values of its derivatives — say an integral
6. $f(x)$ at interest and the values of its derivatives — say an integral

7. $$\int_a^b F(y, y', x)dx, \quad y = f(x).$$

8.

We will use the quote as an example of getting feedback from a type checker. We start by declaring two types, X and Y, and a function f between them:

```
data X   -- X must include the interval [a, b] of the reals
data Y   -- another subset of the reals
f :: X → Y
f = undefined
```

To the Haskell interpreter, such empty **data**-declarations mean that there is no way to construct any element for them, as we saw in Section 2.1.6. But at this stage of the specification, we will use this notation to indicate that we do not know anything about values of those types. Similarly, f has a type, but no proper implementation. We will declare types for the rest of the variables as well, and because we are not implementing any of them right now, we can just make provide a dummy definition for a few of them in one go:

$$(x, deriv, f\!f, a, b, int) = undefined$$

We write $f\!f$ for the capital F (to satisfy Haskell rules for variable names), *deriv* for the derivation operator (D above), and *int* for the integral operator. On line 1 "values of x" hints at the type X for x and the way y is used indicates that it is to be seen as an alias for f (and thus must have the same type).

As we have discussed above, the derivative normally preserves the type and thus we can write:

```
x :: X
y :: X → Y; y = f
y' :: X → Y; y' = deriv f
deriv :: (X → Y) → (X → Y)
```

Next up (on line 5) is the "expression *F*" (which we write *ff*). It should take
three arguments: y, y', x, and return "a quantity". We can invent a new type Z
and write:

> **data** Z -- Probably also some subset of the real numbers
> $ff :: (X \to Y) \to (X \to Y) \to X \to Z$

Then we have the operation of definite integration, which we know should
take two limits $a, b :: X$ and a function $X \to Z$. The traditional mathematics
notation for integration uses an expression (in x) followed by dx, but we can
treat that as a function *expr* binding x:

> $a, b :: X$
> *integral* $= int\ a\ b\ expr$
>   **where** *expr* $x = ff\ y\ y'\ x$
> $int :: X \to X \to (X \to Z) \to Z$

Now we have reached a stage where all the operations have types and pthe
type-checker is happy with them. At this point it is possible to experiment
with variations based on alternative interpretations of the text. For this kind
of "refactoring" is very helpful to have the type checker to make sure the types
still make sense. For example, we could write $ff2 :: Y \to Y \to X \to Z$ as a
variant of *ff* as long as we also change the expression in the integral:

> $ff2 :: Y \to Y \to X \to Z$
> $ff2 = undefined$
> *integral2* $= int\ a\ b\ expr$
>   **where** *expr* $x = ff2\ y\ y'\ x$
>   **where** $y\ = f\ x$
> $y' = deriv\ f\ x$

Both versions (and a few more minor variations) would be fine as exam solutions, but something where the types don't match up would not be OK.

The kind of type inference we presented so far in this chapter becomes automatic with experience in a domain, but is very useful in the beginning.

3.5   Type classes

One difficulty when reading (and implementing) mathematics is *overloading*.
For our purposes, we say that a symbol is *overloaded* when its meaning depends on the type of the expressions that it applies to.

Consider, for example, the operator $(+)$. According to usual mathematical notation, one can use it to add integers, rational numbers, real numbers, complex numbers, etc. and it poses no difficulty. We explore the mathematical reasons in more detail in Section 4.2.2, but for now we will concentrate on the view of functional programming of this problem: one way to understand overloading is via *type classes*.

In Haskell both 4 == 3 and 3.4 == 3.2 typecheck because both integers and floating point values are member of the *Eq* class, which we can safely assume to be defined as follows:

> **class** *Eq a* **where** $(==) :: a \rightarrow a \rightarrow Bool$

The above declaration does two things. First, it names a set of types which have equality test. One can tell the Haskell compiler that certain types belong to this set by using instance declarations, which additionally provide an implementation for the equality test. For example, we can make *Bool* member of the *Eq* using the following declaration:

> *eqBool* :: *Bool* $\rightarrow$ *Bool* $\rightarrow$ *Bool*
> *eqBool True True* = *True*
> *eqBool False False* = *True*
> *eqBool* _ _ = *False*
> **instance** *Eq Bool* **where** $(==) = eqBool$

(The Haskell compiler will in fact provide instances for primitive types).

Second, the *Eq* class declaration provides an operator $(==)$ of type *Eq a* $\Rightarrow$ *a* $\rightarrow$ *a* $\rightarrow$ *Bool*. One can use the operator on any type *a* which belongs to the *Eq* set. This is expressed in general by a constraint *Eq a* occurring before the $\Rightarrow$ symbol.

Instance declarations can also be parameterised on another instance. Consider for example:

> **instance** *Eq a* $\Rightarrow$ *Eq* $[a]$ **where** $(==) = \ldots$ -- exercise

In the above, the expression *Eq a* $\Rightarrow$ *Eq* $[a]$ means that for any type *a* which is already an instance of *Eq* we also make the type $[a]$ an instance of Eq. Thus, for example, by recursion we now have an infinite collection of instances of *Eq*: *Char*, $[Char]$, $[[Char]]$, etc.

3.5.1 Numeric operations

Haskell also provides a *Num* class, containing various numeric types (*Int*, *Double*, etc.) with several operators ($+, *$, etc.). Unfortunately, the *Num* class was designed with more regard for implementation quirks than mathematical structure, and thus it is a poor choice for us. We take a more principled approach instead, and define the following classes, which together serve a similar role as *Num*, and which we study in more detail in Section 4.1:

> **class** *Additive a* **where**
>   *zero* :: *a*
> $(+) :: a \to a \to a$
> **class** *Additive a* $\Rightarrow$ *AddGroup a* **where**
>   *negate* :: $a \to a$ -- specified as $x + negate\ x$ == *zero*
> **class** *Multiplicative a* **where**
>   *one* :: *a*
> $(*) :: a \to a \to a$
> **class** *Multiplicative a* $\Rightarrow$ *MulGroup a* **where**
>   *recip* :: $a \to a$ -- reciprocal, specified as $x * recip\ x$ == *one*

The operator names clash with the *Num* class, which we will avoid from now on in favour *Additive* and *Multiplicative*. In Section 4.1 we will get back to these classes and present a comparison in Fig. 4.1.

Exercise 3.2. Consider the exponentiation operator, which we can write $(\hat{\ })$. Taking advantage of the above classes, propose a possible type for it and sketch an implementation.

Solution: One possibility is $(\hat{\ }) :: MulGroup\ a \Rightarrow a \to Int \to a$. For positive exponents, one can use repeated multiplication. For negative exponents, one can use repeated division.

3.5.2 Overloaded integer literals

We will spend some time explaining a convenient Haskell-specific syntactic shorthand which is very useful but which can be confusing: overloaded integers. In Haskell, every use of an integer literal like 2, 1738, etc., is actually implicitly an application of *fromInteger* to the literal typed as an *Integer*.

But what is *fromInteger*? It is a function that converts integers to any type that supports *zero*, *one*, $(+)$, and $(-)$. We can implement it by the following three cases depending on the sign of n:

$$fromInteger :: (AddGroup\ a, Multiplicative\ a) \Rightarrow Integer \to a$$
$$
\begin{aligned}
fromInteger\ n\ &|\ n < 0 &&= negate\ (fromInteger\ (negate\ n)) \\
&|\ n == 0 &&= zero \\
&|\ otherwise &&= one + fromInteger\ (n - 1)
\end{aligned}
$$

Exercise 3.3. Define *fromRational* which does the same but also handles rational numbers and has the *MulGroup a* constraint.

This means that the same program text can have various meanings depending on the type of the context (but see also Exercise 4.4): The literal *three* = 3, for example, can be used as an integer, a real number, a complex number, or anything which belongs both to *AddGroup* and *Multiplicative*.

3.5.3 Structuring DSLs around type classes

Type classes are related to mathematical structures which, in turn, are related to DSLs. As an example, consider again the DSL of expressions of one variables. We saw that such expressions can be represented by the type $\mathbb{R} \to \mathbb{R}$ in the shallow embedding. Using type classes, we can use the usual operators names instead of *funAdd*, *funMul*, etc. We could write:

> **instance** *Additive* $(\mathbb{R} \to \mathbb{R})$ **where**
> $(+) = funAdd$
> $zero = funConst\ zero$

The instance declaration of the method *zero* above looks recursive, but is not: *zero* is used at a different type on the left- and right-hand-side of the equal sign, and thus refers to two different definitions. One the left-hand-side we define $zero :: \mathbb{R} \to \mathbb{R}$, while on the right-hand-side we use $zero :: \mathbb{R}$.

However, as one may suspect, for functions, we can use any domain and any numeric co-domain in place of $\mathbb{R}$. Therefore we prefer to define the more general instances in Fig. 3.1. Here we extend our set of type-classes to cover algebraic and transcendental numbers. A simplified version, which is sufficient for our purposes, looks as follows:

> **class** *Field a* $\Rightarrow$ *Algebraic a* **where**
> $\sqrt{\cdot} :: a \to a$
> **class** *Field a* $\Rightarrow$ *Transcendental a* **where**
> $\pi :: a$
> $exp :: a \to a$
> $sin :: a \to a$
> $cos :: a \to a$

While classes up to *Field* follow mathematical conventions very closely, for *Algebraic* and *Transcendental* we take the pragmatic approach and list only the methods which are necessary for our development.

instance *Additive a* $\Rightarrow$ *Additive* $(x \to a)$ **where**
  $(+)$ $= funAdd$
  zero $= funConst\ zero$
instance *Multiplicative a* $\Rightarrow$ *Multiplicative* $(x \to a)$ **where**
  $(*)$ $= funMul$
  one $= funConst\ one$
instance *AddGroup a* $\Rightarrow$ *AddGroup* $(x \to a)$ **where**
  $negate\ f = negate \circ f$
instance *MulGroup a* $\Rightarrow$ *MulGroup* $(x \to a)$ **where**
  $recip\ f\ \ = recip \circ f$
instance *Algebraic a* $\Rightarrow$ *Algebraic* $(x \to a)$ **where**
  $\sqrt{f}\ \ \ \ = \sqrt{\cdot} \circ f$
instance *Transcendental a* $\Rightarrow$ *Transcendental* $(x \to a)$ **where**
  $\pi = const\ \pi$
  $sin\ f = sin \circ f;\quad cos\ f = cos \circ f;\quad exp\ f = exp \circ f$

Figure 3.1: Numeric instances lifted to functions. Full definitions can be found in **module** *Algebra* in the repo. (Sometimes referred to as *FunNumInst*.)

Together, these type classes represent an abstract language of abstract and standard operations, abstract in the sense that the exact nature of the elements involved is not important from the point of view of the type class, only from that of its implementation. What does matter for the class (but is not captured in the Haskell definition of the class), is the relationship between various operations (for example addition should distribute over multiplication).

These instances for functions allow us to write expressions which are very commonly used in maths books, such as $f + g$ for the sum of two functions f and g, say $sin + cos :: \mathbb{R} \to \mathbb{R}$. Somewhat less common notations, like $sq *$ $double :: \mathbb{Z} \to \mathbb{Z}$ are also possible. They have a consistent meaning: the same argument is passed to all functions in an expression. As another example, we can write $sin\hat{}2$, which the above instance assigns the following meaning:

$$sin\hat{}2 = \lambda x \to (sin\ x)\hat{}(const\ 2\ x) = \lambda x \to (sin\ x)\hat{}2$$

thus the typical maths notation sin^2 can work fine in Haskell, provided the above instances for functions, assuming a fixed argument. (Note that there is a clash with another common use of superscript for functions in mathematical texts: sometimes $f\hat{}n$ means *composition* of f with itself n times. With that reading sin^2 would mean $\lambda x \to sin\ (sin\ x)$.)

Exercise 3.4. Experiment with this feature using GHCi, for example by evaluating *sin* + *cos* at various points.

Something which may not be immediately obvious, but is nonetheless useful, is that all the above instances are of the form $C\ a\ \Rightarrow\ C\ (x\ \rightarrow\ a)$ and are therefore parametric. This means that, for example, given the instance *Additive* $a \Rightarrow$ *Additive* $(x \rightarrow a)$ and the instance *Additive* $\mathbb{R}$, we have that the types $a \rightarrow \mathbb{R}, a \rightarrow (b \rightarrow \mathbb{R})$, etc. are all instances of *Additive*. Consequently, we can use the usual mathematical operators for functions taking any number of arguments — provided that they match in number and types.

3.6   Computing derivatives

An important part of calculus is the collection of laws, or rules, for computing derivatives. They are provided by Adams and Essex [2010] as a series of theorems, starting at page 108 of their book. We we can summarise those as follows:

$$(f + g)'(x) = f'(x) + g'(x)$$
$$(f * g)'(x) = f'(x) * g(x) + f(x) * g'(x)$$
$$(C * f)'(x) = C * f'(x)$$
$$(f \circ g)'(x) = f'(g(x)) * g'(x) \quad \text{-- chain rule}$$

(After a while, Adams and Essex switch to differential notation, so we omit corresponding rules for trigonometric and exponential functions.) Using the notation $D\ f$ for the derivative of f and applying the numeric operations to functions directly, we can fill in a table of examples which can be followed to compute derivatives of many functions:

$$
\begin{aligned}
D\ (f + g) &= D\,f + D\,g \\
D\ (f * g) &= D\,f * g + f * D\,g \\
D\ id &= const\ 1 \\
D\ (const\ a) &= const\ 0 \\
D\ (f \circ g) &= (D\,f \circ g) * D\,g \quad \text{-- chain rule} \\
D\ sin &= cos \\
D\ cos &= -sin \\
D\ exp &= exp
\end{aligned}
$$

and so on.

If we want to get a bit closer to actually implementing D we quickly notice a problem: if D has type $(\mathbb{R} \rightarrow \mathbb{R}) \rightarrow (\mathbb{R} \rightarrow \mathbb{R})$, we have no way to turn the above specification into a program, because the program has no way of telling which of these rules should be applied. That is, given an extensional (semantic, shallow) function f, the only thing that we can ever do is to evaluate f at given points, and thus we cannot know if this function was written using a $(+)$, or sin or exp as outermost operation. The only thing that a derivative operator could do would be to numerically approximate the derivative, and that is not what we are exploring in this book. Thus we need to take a step back and change the type that we work on. Even though the rules in the table are obtained by reasoning semantically, using the definition of limit for functions (of type $\mathbb{R} \to \mathbb{R}$), they are really intended to be used on *syntactic* functions or expressions: abstract syntax trees *representing* the (semantic) functions.

We observe that we can compute derivatives for any expression made out of arithmetic functions, trigonometric functions, the exponential and their compositions. In other words, the computation of derivatives is based on a domain specific language of expressions (representing functions in one variable). This means that we can in fact implement the derivative of *FunExp* expressions (from Section 1.7.1), using the rules of derivatives. Because the specification of derivation rules is already in the right format, the way to obtain this implementation may seem obvious, but we will go through the steps as a way to show the process in a simple case.

Our goal is to implement a function $derive :: FunExp \to FunExp$ which makes the following diagram commute:

$$FunExp \xrightarrow{\;eval\;} Func$$
$$\Big\downarrow derive \qquad\qquad \Big\downarrow D$$
$$FunExp \xrightarrow{\;eval\;} Func$$

That is, we want the following equality to hold:

$$eval \circ derive = D \circ eval$$

In turn, this means that for any expression $e :: FunExp$, we want

$$eval\ (derive\ e) = D\ (eval\ e)$$

As an example of using equational reasoning we will calculate the definition of *derive* for a specific constructor. We have added $Exp :: FunExp \to FunExp$ to the datatype *FunExp* for this example. We start from the left-hand side of the specification, in the case when the expression is of the form $Exp\ e$, and use equalities from the mathematical side in combination with our instances and functions definitions to "push" the call of *derive* to the subexpression e:

$$
\begin{aligned}
&eval\ (derive\ (Exp\ e)) && = \{\text{- specification of } derive \text{ above -}\} \\
&D\ (eval\ (Exp\ e)) && = \{\text{- def. } eval \text{ -}\} \\
&D\ (exp\ (eval\ e)) && = \{\text{- def. } exp \text{ for functions -}\} \\
&D\ (exp \circ eval\ e) && = \{\text{- chain rule -}\} \\
&(D\ exp \circ eval\ e) * D\ (eval\ e) && = \{\text{- } D \text{ rule for } exp \text{ -}\} \\
&(exp \circ eval\ e) * D\ (eval\ e) && = \{\text{- specification of } derive \text{ -}\} \\
&(exp \circ eval\ e) * (eval\ (derive\ e)) && = \{\text{- def. of } eval \text{ for } Exp \text{ -}\} \\
&(eval\ (Exp\ e)) * (eval\ (derive\ e)) && = \{\text{- def. of } eval \text{ for } :*: \text{ -}\} \\
&eval\ (Exp\ e\ :*:\ derive\ e)
\end{aligned}
$$

Therefore, the specification is fulfilled by taking

$$derive\ (Exp\ e) = Exp\ e\ :*:\ derive\ e$$

Similarly, we obtain

$$
\begin{aligned}
&derive :: FunExp \rightarrow FunExp \\
&derive\ (Const\ \alpha) && = Const\ 0 \\
&derive\ X && = Const\ 1 \\
&derive\ (e_1\ :+:\ e_2) && = derive\ e_1\ :+:\ derive\ e_2 \\
&derive\ (e_1\ :*:\ e_2) && = (derive\ e_1\ :*:\ e_2)\ :+:\ (e_1\ :*:\ derive\ e_2) \\
&derive\ (Exp\ e) && = Exp\ e\ :*:\ derive\ e
\end{aligned}
$$

Exercise 3.5. Complete the *FunExp* type and the *eval* and *derive* functions.

3.7   Exercises

Exercise 3.6. To get a feeling for the Lagrange equations, let us take the case of an object falling in constant gravity: $L\ (t, q, v) = m * v\hat{}2\ /\ 2 - m * g * q$, compute *expand w*, perform the derivatives and check if the equation is satisfied for the following three cases. If, in one of the cases, the equation is not satisfied in general, see if you can find some values of the mass m and the acceleration due to gravity g which makes the equations hold.

- $w_1 = id$ or

- $w_2 = sin$ or

- $w_3 = (q0-) \circ (g*) \circ (/2) \circ (\hat{}2)$

Exercise 3.7. Consider the following text from Mac Lane's *Mathematics Form and Function* (page 168):

> If $z = g(y)$ and $y = h(x)$ are two functions with continuous derivatives, then in the relevant range $z = g(h(x))$ is a function of x and has derivative
> $$z'(x) = g'(y) * h'(x)$$

Give the types of the elements involved ($x, y, z, g, h, z', g', h', *$ and $'$).

Exercise 3.8. Consider the following text from Mac Lane's *Mathematics Form and Function* (page 182):

> In these cases one tries to find not the values of x which make a given function $y = f(x)$ a minimum, but the values of a given function $f(x)$ which make a given quantity a minimum. Typically, that quantity is usually measured by an integral whose integrand is some expression F involving both x, values of the function $y = f(x)$ at interest and the values of its derivatives – say an integral
> $$\int_a^b F(y, y', x)dx, \quad y = f(x).$$

Give the types of the variables involved (x, y, y', f, F, a, b) and the type of the four-argument integration operator:

$$\int_{.}^{.} \cdot d\cdot$$

Exercise 3.9. In the case of the simplest probability theory, we start with a *finite*, non-empty set Ω of *elementary events*. An *event* is a subset of Ω, i.e. an element of the powerset of Ω, (that is, $\mathcal{P}\,\Omega$). A *probability function* P associates to each event a real number between 0 and 1, such that

- $P\,\varnothing = 0, P\,\Omega = 1$

- A and B are disjoint (i.e., $A \cap B = \varnothing$), then: $P\,A + P\,B = P\,(A \cup B)$.

Chapter 9 presents more material about probability theory, but that material is not needed for the present exercise.

Conditional probabilities are defined as follows [Stirzaker, 2003]:

> Let A and B be events with $P\,B > 0$. given that B occurs, the *conditional probability* that A occurs is denoted by $P\,(A \mid B)$ and defined by $P\,(A \mid B) = P\,(A \cap B)\,/\,P\,B$

1. What are the types of the elements involved in the definition of conditional probability?
   ($P, \cap, /, \mid$)

2. In the 1933 monograph that set the foundations of contemporary probability theory, Kolmogorov used, instead of $P\,(A \mid B)$, the expression $P_B\,A$. Type this expression. Which notation do you prefer (provide a *brief* explanation).

Exercise 3.10. Multiplication for matrices (from the matrix algebra DSL).

Consider the following definition, from "Linear Algebra" by Donald H. Pelletier:

Definition: If A is an $m \times n$ matrix and B is an $n \times p$ matrix, then the *product*, AB, is an $m \times p$ matrix; the $(i, j)^{th}$ entry of AB is the sum of the products of the pairs that are obtained when the entries from the ith row of the left factor, A, are paired with those from the jth column of the right factor, B.

1. Introduce precise types for the variables involved: A, m, n, B, p, i, j. You can write *Fin n* for the type of the values $\{0, 1, \ldots, n-1\}$.

2. Introduce types for the functions *mul* and *proj* where $AB = mul\ A\ B$ and *proj i j M* = "take the $(i, j)^{th}$ entry of M". What class constraints (if any) are needed on the type of the matrix entries in the two cases?

3. Implement *mul* in Haskell. You may use the functions *row* and *col* specified by *row* i M = "the ith row of M" and *col* j M = "the jth column of M". You don't need to implement them and here you can assume they return plain Haskell lists.

Exercise 3.11. (*Extra material outside the course.) In the same direction as the Lagrangian case study in Section 3.3 there are two nice blog posts by Justin Le about Hamiltonian dynamics implemented in Haskell: one introductory and one more advanced. It is a good exercise to work through the examples in these posts.

Chapter 4

Compositionality and Algebras

Algebraic structures are fundamental to the structuralist point of view in mathematics, which emphasises relations between objects rather than the objects themselves and their representations. Furthermore, each mathematical domain has its own fundamental structures. Once these structures have been identified, one tries to push their study as far as possible *on their own terms*, without picking any particular representation (which may have richer structure than the one we want to study). For example, in group theory, one starts by exploring the consequences of just the group structure, rather than introducing any particular group (like integers) which have (among others) an order structure and monotonicity.

Furthermore, mappings or (translations) between such structures become an important topic of study. When such mappings preserve the structure, they are called *homomorphisms*. As two examples, we have the homomorphisms *exp* and *log*, specified as follows:

$$
\begin{aligned}
&exp : \mathbb{R} &&\to \mathbb{R}_{>0} \\
&exp\ 0 &&= 1 &&\text{-- } e^0 = 1 \\
&exp\ (a+b) &&= exp\ a * exp\ b &&\text{-- } e^{a+b} = e^a e^b \\
&log : \mathbb{R}_{>0} &&\to \mathbb{R} \\
&log\ 1 &&= 0 &&\text{-- } \log 1 = 0 \\
&log\ (a*b) &&= log\ a + log\ b &&\text{-- } \log(ab) = \log a + \log b
\end{aligned}
$$

What we recognise as the familiar laws of exponentiation and logarithms arise

101

from homomorphism conditions, which relate the additive and multiplicative structures of reals and positive reals.

Additionally, homomorphisms play a crucial role when relating an abstract syntax (a datatype), and a semantic domain (another type) via an evaluation function between them (the semantics). In this chapter we will explain the notions of algebraic structure and homomorphism in detail and show applications both in mathematics and DSLs in general.

4.1   Algebraic Structures

What is an algebraic structure? Let's turn to Wikipedia (Universal algebra, 2020-09-25) as a starting point:

> In universal algebra, an algebra (or algebraic structure) is a set A together with a collection of operations on A (of finite arity) and a collection of axioms which those operation must satisfy.

The fact that a type a is equipped with operations is conveniently captured in Haskell using a type class (Section 3.5).

Example   A particularly pervasive structure is that of monoids. A monoid is an algebra which has an associative operation *op* and a *unit*:

```
class Monoid a where
    unit :: a
    op   :: a → a → a
```

The laws cannot easily be captured in the Haskell class, but can be formulated as the following propositions:

$$\forall x : a. \ (unit \ `op` \ x \ == \ x \land x \ `op` \ unit \ == \ x)$$
$$\forall x, y, z : a. \ (x \ `op` \ (y \ `op` \ z) \ == \ (x \ `op` \ y) \ `op` \ z)$$

The first law ensures that *unit* is indeed the unit of *op* and the second law is the familiar associativity law for *op*.

Example   Examples of monoids include numbers with addition, $(\mathbb{R}, 0, (+))$, positive numbers with multiplication $(\mathbb{R}_{>0}, 1, (*))$, and even endofunctions with composition $(a → a, id, (\circ))$. An *endofunction* is a function of type $X → X$ for some set X. An endofunction which also preserves structure is called an endomorphism.

Exercise 4.1. Define the above monoids as type class instances and check that the laws are satisfied.

Example To make this concept a bit more concrete, we define two examples in Haskell: the additive monoid $\mathbb{N}_A$ and the multiplicative monoid $\mathbb{N}_M$.

newtype $\mathbb{N}_A$ $= A\ \mathbb{N}$ **deriving** $(Show, Eq)$
instance *Monoid* $\mathbb{N}_A$ **where**
    unit $= A\ zero$
    op $(A\ m)\ (A\ n)\ = A\ (m + n)$
newtype $\mathbb{N}_M$ $= M\ \mathbb{N}$ **deriving** $(Show, Eq)$
instance *Monoid* $\mathbb{N}_M$ **where**
    unit $= M\ one$
    op $(M\ m)\ (M\ n) = M\ (m * n)$

In Haskell there can be at most one instance of a given class in scope for a given type, so we cannot define two **instance** *Monoid* $\mathbb{N}$: we must make a **newtype** whose role is to indicate which of the two possible monoids (additive or multiplicative) applies in a given context. But, in mathematical texts the constructors M and A are usually omitted, and instead the names of the operations suggest which of the monoids one is referring to. To be able to conform to that tradition we defined two separate classes in Section 3.5.1, one for the additive and one for the multiplicative monoids, as follows:

class *Additive a* **where** *zero* $:: a$; $(+) :: a \to a \to a$
class *Multiplicative a* **where** *one* $:: a$; $(*) :: a \to a \to a$

4.1.1 Groups and rings

Another important structure are groups, which are monoids augmented with an inverse. To continue our mathematically-grounded *Num* replacement, we have also defined the additive group as follows:

class *Additive a* $\Rightarrow$ *AddGroup a* **where**
    negate $:: a \to a$

Groups demand that the inverse (called *negate* for the additive group) act like an inverse: applying the operation to an element and its inverse should yield the unit of the group. Thus, for the additive group, the laws are:

$\forall a.\ negate\ a + a = zero$
$\forall a.\ a + negate\ a = zero$

With *negate* (for "unary minus") in place we can define subtraction as

$$(-) :: AddGroup\ a \Rightarrow a \rightarrow a \rightarrow a$$
$$a - b = a + negate\ b$$

Note that this definition works for any (additive) group, thus the type a need not represent numbers. Just with instances from this book we can subtract functions, pairs, power series, etc.

When the additive monoid is abelian (commutative) and multiplication distributes over addition ($x * (y + z)\ ==\ (x * y) + (x * z)$), we have a *Ring*. As always we cannot conveniently specify laws in Haskell type classes and thus define *Ring* simply as the conjunction of *AddGroup* and *Multiplicative*:

type $Ring\ a = (AddGroup\ a, Multiplicative\ a)$

With that, we have completed the structural motivation of our replacement for the *Num* class!

Exercise 4.2. Prove that $\mathbb{N}$ admits the usual *Additive* instance. Likewise for *Ring* instances of $\mathbb{Z}$, $\mathbb{Q}$, and $\mathbb{R}$.

We note right away that one can have a multiplicative group structure as well, whose inverse is called the reciprocal (abbreviated as *recip* in Haskell). With that in place, division can be defined in terms of multiplication and reciprocal.

class $Multiplicative\ a \Rightarrow MulGroup\ a$ **where**
    $recip :: a \rightarrow a$ -- reciprocal
$(/) :: MulGroup\ a \Rightarrow a \rightarrow a \rightarrow a$
$a\ /\ b = a * recip\ b$

Often the multiplicative group structure is added to a ring to get what is called a *field* which we represent by the type class *Field*:

type $Field\ a = (Ring\ a, MulGroup\ a)$

For fields, the reciprocal is not defined at zero. We will not capture this precondition in types: it would cause too much notational awkwardness. Example instances of *Field* are $\mathbb{Q}$ and $\mathbb{R}$. For pragmatic reasons we will also treat *Double* as a *Field* even though the laws only hold approximately.

In Fig. 4.1 the numerical classes we use in this book are summed up and compared with the Haskell *Num* class hierarchy.

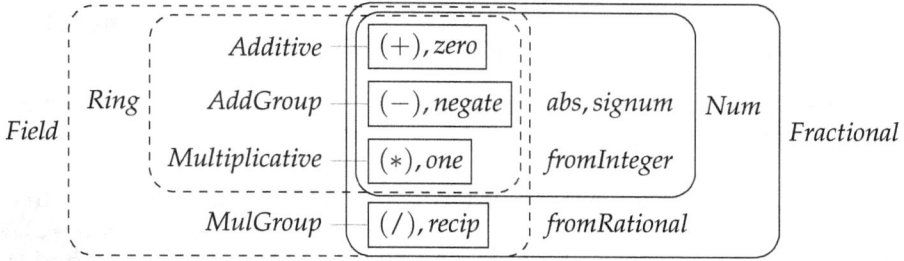

Figure 4.1: Comparing the Haskell Prelude class hierarchy (*Num, Fractional*) with the book's hierarchy. In addition to the groupings visible in the figure, the class *AddGroup* includes the *Additive* operations and *MulGroup* includes the *Multiplicative* operations.

4.2 Homomorphisms

The Wikipedia definition of homomorphism (2021-12-27) states that "A homomorphism is a structure-preserving map between two algebraic structures of the same type". We will spend the next few subsections on formal definitions and examples of homomorphisms.

4.2.1 (Homo)morphism on one operation

As a stepping stone to capture the idea of homomorphism, we can define a ternary predicate H_2. The first argument h, is the map. The second (Op) and third (op) arguments correspond to the algebraic structures.

$$H_2(h, Op, op) = \forall\, x.\ \forall\, y.\ h\ (Op\ x\ y) == op\ (h\ x)\ (h\ y)$$

If the predicate $H_2(h, Op, op)$ holds, we say that $h : A \to B$ is a homomorphism from $Op : A \to A \to A$ to $op : B \to B \to B$. Or that h is a homomorphism from Op to op. Or even that h is a homomorphism from A to B if the operators are clear from the context. We have seen several examples in earlier chapters:

1. in Section 1.4 we saw that *evalE* : *ComplexE* $\to$ *ComplexD* is a homomorphism from the syntactic operator *Plus* to the corresponding semantic operator *plusD*.

2. in Chapter 2 we saw De Morgan's laws, which say that "not" ($\neg$) is a homomorphism in two ways: $H_2(\neg, (\wedge), (\vee))$ and $H_2(\neg, (\vee), (\wedge))$.

3. in Section 1.7.1 we saw that $eval : FunExp \rightarrow Func$ is a homomorphism from syntactic $(:*:)$ to semantic $(*)$ for functions

4. If $(*)$ distributes over $(+)$ for some type A then $(*c) : A \rightarrow A$ is a homomorphism from $(+)$ to $(+)$: $H_2((*c), (+), (+))$.

To see how this last item plays out, it can be helpful to study the syntax trees of the left and right hand sides of the distributive law: $((a + b) * c = (a * c) + (b * c))$. We observe that the function $(*c)$ is "pushed down" to both a and b:

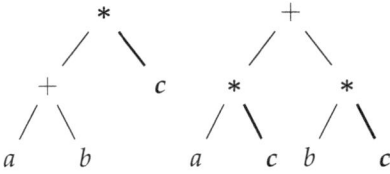

Exercise 4.3. Expand the definition of H_2 in each case and check that the obtained conditions hold.

4.2.2   Homomorphism on structures

So far our definition of homomorphism takes the rather limited view that a single operation is transformed. Usually, homomorphisms map a whole *structure* (or *algebra*) with several operations.

Back to Wikipedia:

> More formally, a homomorphism between two algebras A and B is a function $h : A \rightarrow B$ from the set A to the set B such that, for every operation f_A of A and corresponding f_B of B (of arity, say, n), $h (f_A (x_1, \ldots, x_n)) = f_B (h (x_1), \ldots, h (x_n))$.

In our Haskell interpretation, the above would mean that we have $H_2(h, f_A, f_B)$ for every binary method f in a given class C and more generally $H_n (h, op_A, op_B)$ for each operation op of arity n. We can also use type class overloading to write $H_n (h, op, op)$ where the first occurrence of op comes from the C A instance and the second one from C B.

Example   The general monoid homomorphism conditions for $h : A \rightarrow B$ are:

```
h unit     = unit        -- h takes units to units
h (x 'op' y) = h x 'op' h y   -- and distributes over op (for all x and y)
```

Note that both *unit* and *op* have different types on the left and right hand sides. On the left they belong to the monoid $(A, unit_A, op_A)$ and on the right the belong to $(B, unit_B, op_B)$.

Example Hence, the function *exp* is a monoid homomorphism from $(\mathbb{R},0,(+))$ to $(\mathbb{R}_{>0},1,(*))$.

$$exp : \mathbb{R} \quad \rightarrow \mathbb{R}_{>0}$$
$$exp\ 0 \quad = 1 \qquad\qquad -\!\!- e^0 = 1$$
$$exp\ (a+b) = exp\ a * exp\ b \quad -\!\!- e^{a+b} = e^a e^b$$

In the above example, we have simply checked the homomorphism conditions for the exponential function. But we can try to go the other way around: knowing that a function h is homomorphism, what kind of function can h be?

Example Let us characterise the homomorphisms from $\mathbb{N}_A$ to $\mathbb{N}_M$ (from Section 4.1). What can be inferred from just the homomorphism conditions?

Let $h : \mathbb{N}_A \rightarrow \mathbb{N}_M$ be a monoid homomorphism. Then it must satisfy the following conditions:

$$h\ 0 \quad\quad = 1$$
$$h\ (x+y) = h\ x * h\ y \quad -\!\!- \text{for all } x \text{ and } y$$

For example $h\ (x+x) = h\ x * h\ x = (h\ x)\hat{\ }2$ which for $x = 1$ means that $h\ 2 = h\ (1+1) = (h\ 1)\hat{\ }2$.

More generally, every natural number is equal to the sum of n ones: $1 + 1 + \dots + 1$. Therefore

$$h\ n = h\ (1 + \dots + 1)$$
$$= h\ 1\ * \dots\ * h\ 1$$
$$= (h\ 1)\hat{\ }n$$

That is, every choice of $h\ 1$ induces a homomorphism from $\mathbb{N}_A$ to $\mathbb{N}_M$. This means that the value of the function h, for any natural number, is fully determined by its value for 1.

In other words, we know that every h (homomorphism from $\mathbb{N}_A$ to $\mathbb{N}_M$) is of the form

$$h\ n = a\hat{\ }n$$

for a given natural number $a = h\ 1$. So, the set of homomorphisms between the additive monoid and the multiplicative monoid is the set of exponential functions, one for every base a. Note how almost all functions from $\mathbb{N}$ to $\mathbb{N}$ are ruled out by the homomorphism conditions.

Exercise 4.4. Assume an arbitrary *Ring*-homomorphism f from *Integer* to an arbitrary type a. Prove f == *fromInteger*, provided the definition in Section 3.5.2.

Solution: The homomorphism conditions include:

$$
\begin{array}{ll}
f\ zero & \texttt{==}\ zero & \text{-- from } H_0(f, zero, zero) \\
f\ (one + x) & \texttt{==}\ one + f\ x & \text{-- from } H_2(f, (+), (+)) \\
f\ (negate\ x) & \texttt{==}\ negate\ (f\ x) & \text{-- from } H_1(f, negate, negate)
\end{array}
$$

By substitution we get the following equations:

$$
\begin{array}{ll}
f\ zero & \texttt{==}\ zero \\
f\ x & \texttt{==}\ one + (f\ (x - one)) \\
f\ x & \texttt{==}\ negate\ (f\ (negate\ x))
\end{array}
$$

These are compatible with the behaviour of *fromInteger*, but they also completely fix the behaviour of f if x is an integer, because it can either be zero, positive or negative.

Other homomorphisms

Exercise 4.5. Show that *const* is a homomorphism.

Solution: This exercise is underspecified (what structure? from and to which types?) so we need to explore a bit to find a reasonable interpretation. We can start simple and use addition $(+)$ as the structure, thus we want to show $H_2(h, (+), (+))$ where $h = const$ is of type $A \to B$. Next we need to identify the types A and B where addition is used in the predicate. We have $const ::$ $a \to (x \to a)$ for any types a and x and we can take $A = a = \mathbb{R}$ and $B = x \to \mathbb{R}$. As B is a function type the $(+)$ on that side is addition of functions, which we defined in Section 3.5.3 in terms of *funAdd* from Section 1.7.1. The homomorphism law (that h distributes over $(+)$) can be shown as follows:

$$
\begin{array}{ll}
h\ (a + b) & = \{\text{- } h = const \text{ in this case -}\} \\
const\ (a + b) & = \{\text{- By def. of } const \text{ -}\} \\
(\lambda x \to a + b) & = \{\text{- By def. of } const, \text{ twice, backwards -}\} \\
(\lambda x \to const\ a\ x + const\ b\ x) & = \{\text{- By def. of } funAdd, \text{ backwards -}\} \\
funAdd\ (const\ a)\ (const\ b) & = \{\text{- By def. of } (+) \text{ on functions -}\} \\
const\ a + const\ b & = \{\text{- } h = const, \text{ twice -}\} \\
h\ a + h\ b
\end{array}
$$

We now have a homomorphism from values to functions, and you may wonder if there is a homomorphism in the other direction. The answer is "Yes, many". Such homomorphisms take the form *apply c*, for any c.

Exercise 4.6. Show that *apply c* is an *Additive* homomorphism for all *c*, where *apply x f = f x*.

Solution: Indeed, writing *h = apply c* for some fixed *c*, we have

$h (f + g) = \{$- def. *apply* -$\}$
$(f + g) c = \{$- def. $(+)$ for functions -$\}$
$f c + g c = \{$- def. *apply* -$\}$
$h f + h g$

and

$h\ zero = \{$- def. *apply* -$\}$
$zero\ c = \{$- def. *zero* for functions -$\}$
$zero$

4.2.3 *Isomorphisms

Two homomorphisms which are inverse of each other define an *isomorphism*. If an isomorphism exist between two sets, we say that they are isomorphic. For example, the exponential and the logarithm witness an isomorphism between $\mathbb{R}_{>0}$ and $\mathbb{R}$.

Exercise 4.7. Show that exponential and logarithm are inverse of each other.

Exercise 4.8. Extend the *exp-log*-isomorphism to relate the methods of the *AddGroup* and *MulGroup* classes.

Exercise 4.9 (Hard.). Sketch the isomorphism between IPC and STLC seen in Section 2.1.6. What are the structures? What are mappings (functions)?

Exercise 4.10. Sketch an isomorphism between pairs of numbers and complex numbers, as suggested in Section 1.4.

4.3 Compositional semantics

The core of compositional semantics is that the meaning (semantics) of an expression is determined by combining the meanings of its subexpressions. Earlier, we have presented several DSLs as a syntax datatype *Syn* connected to a type *Sem* for the semantic values by a compositional semantic function $eval : Syn \to Sem$. In this section we show that compositional functions are homomorphisms and provide some examples with proofs of compositionality (or the opposite).

4.3.1   Compositional functions are homomorphisms

Consider a datatype of very simple integer expressions:

data $E = Add\ E\ E \mid Mul\ E\ E \mid Con\ Integer$ **deriving** Eq
$e_1, e_2 :: E$ $-- 1 + 2 * 3$
$e_1 = Add\ (Con\ 1)\ (Mul\ (Con\ 2)\ (Con\ 3))$ $-- 1 + (2 * 3)$
$e_2 = Mul\ (Add\ (Con\ 1)\ (Con\ 2))\ (Con\ 3)$ $-- (1 + 2) * 3$

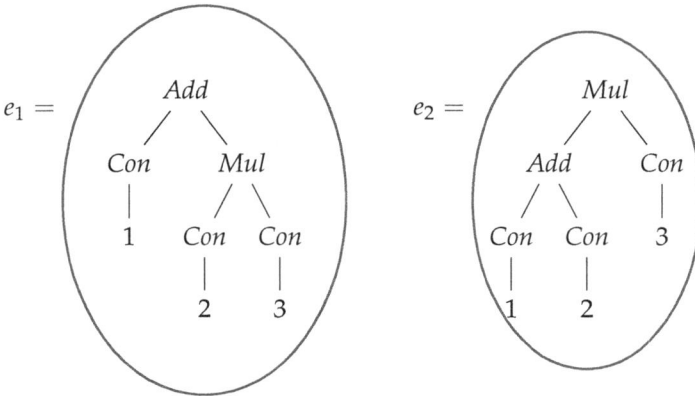

As the reader may have guessed, the natural evaluator $eval : E \rightarrow Integer$ (defined later) is a homomorphism from Add to $(+)$ and from Mul to $(*)$. But to practice the definition of homomorphism we will here check if $even$ or $isPrime$ is a homomorphism from E to $Bool$.

Let's try to define $even : E \rightarrow Bool$ with the usual induction pattern:

$even\ (Add\ x\ y) = evenAdd\ (even\ x)\ (even\ y)$
$even\ (Mul\ x\ y) = evenMul\ (even\ x)\ (even\ y)$
$even\ (Con\ c)\quad = evenCon\ c$

$evenAdd :: Bool \rightarrow Bool \rightarrow Bool$
$evenMul :: Bool \rightarrow Bool \rightarrow Bool$
$evenCon :: Integer \rightarrow Bool$

Note that $even$ throws away lots of information: the domain is infinite and the range is a two-element set. This information loss could make it difficult to define the helper functions $evenAdd$, etc. because they only get to work on the small range. Still, in this case we are lucky: we can use the "parity rules" taught in elementary school: even + even is even, etc. In code we simply get:[1]

---

[1] A perhaps more natural alternative would be to taken odd instead of $even$ as the homomorphism. You can try it out as an exercise.

$$evenAdd = (==)$$
$$evenMul = (\vee)$$
$$evenCon = (0 ==) \circ ('mod'2)$$

Exercise 4.11. Prove $H_2(even, Add, evenAdd)$ and $H_2(even, Mul, evenMul)$.

4.3.2 An example of a non-compositional function

Let's now try to define *isPrime* : $E \rightarrow Bool$ in the same way to see a simple example of a non-compositional function. In this case it is enough to just focus on one of the cases to already see the problem:

isPrimeAdd :: *Bool* $\rightarrow$ *Bool* $\rightarrow$ *Bool*
isPrimeAdd = *error* "Can this be done?"
isPrime (*Add x y*) = *isPrimeAdd* (*isPrime x*) (*isPrime y*)

As before, if we can define *isPrimeAdd*, we will get $H_2(isPrime, Add, isPrimeAdd)$ "by construction". But it is not possible for *isPrime* to both satisfy its specification and $H_2(isPrime, Add, isPrimeAdd)$. To shorten the calculation we write just n for *Con n*.

> *False*
> = {- By spec. of *isPrime* (four is not prime). -}
> *isPrime* (*Add 2 2*)
> = {- by H_2 -}
> *isPrimeAdd* (*isPrime 2*) (*isPrime 2*)
> = {- By spec. of *isPrime* (two is prime). -}
> *isPrimeAdd* (*isPrime 2*) *True*
> = {- By spec. of *isPrime* (three is also prime). -}
> *isPrimeAdd* (*isPrime 2*) (*isPrime 3*)
> = {- by H_2 -}
> *isPrime* (*Add 2 3*)
> = {- By spec. of *isPrime* (five is prime). -}
> *True*

But because we also know that *False* $\neq$ *True*, we have a contradiction. Thus we conclude that *isPrime* is *not* a homomorphism from E to *Bool*, regardless of the choice of the operator (on the Boolean side) corresponding to addition.

4.4 Folds

In general, for a syntax *Syn*, and a possible semantics (a type *Sem* and an *eval* function of type *Syn* $\rightarrow$ *Sem*), we call the semantics *compositional* if we can implement *eval* as a fold. Informally a "fold" is a recursive function which replaces each abstract syntax constructor C_i of *Syn* with its semantic interpretation c_i — but without doing any other change in the structure. In particular, moving around constructors is forbidden. For example, in our datatype E, a compositional semantics means that *Add* maps to *add*, *Mul* $\mapsto$ *mul*, and *Con* $\mapsto$ *con* for some "semantic functions" *add*, *mul*, and *con*.

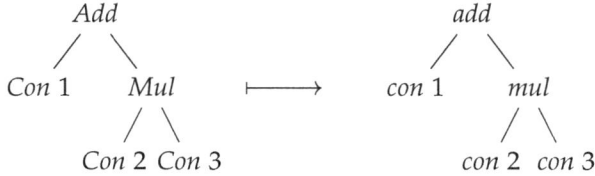

$$
\begin{array}{ccc}
\begin{array}{c}
Add \\
\diagup \ \diagdown \\
Con\ 1 \qquad Mul \\
\diagup \diagdown \\
Con\ 2\ \ Con\ 3
\end{array}
& \longmapsto &
\begin{array}{c}
add \\
\diagup \ \diagdown \\
con\ 1 \qquad mul \\
\diagup \diagdown \\
con\ 2\ \ con\ 3
\end{array}
\end{array}
$$

As an example we can define a general *foldE* for the integer expressions:

$$foldE :: (s \to s \to s) \to (s \to s \to s) \to (Integer \to s) \to (E \to s)$$
foldE add mul con = *rec*
  where *rec* $(Add\ x\ y)$ = *add* $(rec\ x)\ (rec\ y)$
         rec $(Mul\ x\ y)$ = *mul* $(rec\ x)\ (rec\ y)$
         rec $(Con\ i)$ = *con i*

Notice that *foldE* is a higher-order function: it take three function arguments corresponding to the three constructors of E, and returns a function from E to the semantic domain s.

The "natural" evaluator to integers is then easy to define:

$$evalE1 :: E \to Integer$$
evalE1 = *foldE* $(+)\ (*)\ id$

and with a minimal modification we can also make it work for other numeric types:

$$evalE2 :: Ring\ a \Rightarrow E \to a$$
evalE2 = *foldE* $(+)\ (*)\ fromInteger$

Another thing worth noting is that if we replace each abstract syntax constructor with itself we get an identity function, sometimes known as a "deep copy":

$$idE :: E \to E$$
idE = *foldE Add Mul Con*

Finally, it is useful to capture the semantic functions (the parameters to the fold) in a type class:

```
class IntExp t where
    add :: t → t → t
    mul :: t → t → t
    con :: Integer → t
```

In this way we can turn the arguments to the fold into a constraint on the return type:

```
foldIE :: IntExp t ⇒ E → t
foldIE = foldE add mul con
```

We can then provide type class instances for both the natural semantics *Integer* and the syntax *E* and recover the evaluator and the "deep copy" function as special cases:

```
instance IntExp Integer where add = (+); mul = (*); con = id
instance IntExp E       where add = Add; mul = Mul; con = Con

evalE' :: E → Integer
evalE' = foldIE
idE' :: E → E
idE' = foldIE
```

Additionally *IntExp* is the underlying algebraic structure of the fold. The function *foldIE* is a homomorphism which maps the *IntExp E* instance to another (arbitrary) instance *IntExp e*. This is what a fold is in general. Given a structure *C*, a fold is a homomorphism from a realisation of *C* as a data-type. We can note at this point that a class *C a* can be realised as a datatype only if all the functions of **class** *C a* return *a*. (Otherwise the constructors could create another type; and so they are not constructors any more.) This condition was satisfied in the case of our **class** *IntExp t*: all function signatures end with ... → *t*. When this condition is satisfied, we say that the class is an *algebra* — not just any algebraic structure. (Indeed, this terminology can be confusing.)

List-folding The description of a fold also works for the datatype of lists: if we let *Syn* = [*a*] and *Sem* = *s* the constructor (:) :: *a* → *Syn* → *Syn* is replaced by a function *cons* :: *a* → *Sem* → *Sem* and the constructor [ ] :: *Syn* is replaced by *nil* :: *Sem*. Or, if we expand the type synonyms:

```
foldList :: (a → s → s) → s → ([a] → s)
foldList cons nil = rec
    where rec (x : xs) = cons x (rec xs)
          rec []       = nil
```

This is (a special case of) the function called *foldr* in the Haskell prelude.

4.4.1   Even folds can be wrong!

When working with expressions it is often useful to have a "pretty-printer" to convert the abstract syntax trees to strings like "1+2*3".

$$pretty :: E \rightarrow String$$

We can view *pretty* as an alternative *eval* function for a semantics using *String* as the semantic domain instead of the more natural *Integer*. We can implement *pretty* in the usual way as a fold over the syntax tree using one "semantic constructor" for each syntactic constructor:

$$pretty\ (Add\ x\ y) = prettyAdd\ (pretty\ x)\ (pretty\ y)$$
$$pretty\ (Mul\ x\ y) = prettyMul\ (pretty\ x)\ (pretty\ y)$$
$$pretty\ (Con\ c)\quad = prettyCon\ c$$

$$prettyAdd :: String \rightarrow String \rightarrow String$$
$$prettyMul :: String \rightarrow String \rightarrow String$$
$$prettyCon :: Integer \rightarrow String$$

We can also see *String* and *pretty* as an instance of the *IntExp* class:

instance *IntExp String* **where**
   $add = prettyAdd$
   $mul = prettyMul$
   $con = prettyCon$
$pretty' :: E \rightarrow String$
$pretty' = foldIE$

Now, if we try to implement the semantic constructors without thinking too much we would get the following:

$$prettyAdd\ xs\ ys = xs \mathbin{+\!\!+} "+" \mathbin{+\!\!+} ys$$
$$prettyMul\ xs\ ys = xs \mathbin{+\!\!+} "*" \mathbin{+\!\!+} ys$$
$$prettyCon\ c\quad = show\ c$$

$p_1, p_2 :: String$
$p_1 = pretty\ e_1$ -- gives "1+2*3"
$p_2 = pretty\ e_2$ -- also "1+2*3", but should be "(1+2)*3"
$trouble :: Bool$
$trouble = p_1\ \texttt{==}\ p_2$

Note that e_1 and e_2 (from Section 4.3.1) are not equal, but they still pretty-print to the same string. This means that *pretty* is doing something wrong: the inverse, *parse*, is ambiguous. There are many ways to fix this, some more "pretty" than others. One way to characterise the issue is that some information is lost in the translation: *pretty* is not invertible.

Thus, we can see that a function can be a homomorphism and still be "wrong".

For the curious One solution to the problem with parentheses is to create three (slightly) different functions intended for printing in different contexts. The first of them is for the top level, the second for use inside *Add*, and the third for use inside *Mul*. These three functions all have type $E \to String$ and can thus be combined with the tupling transform into one function returning a triple: $prVersions :: E \to (String, String, String)$. The result is the following:

$$prTop :: E \to String$$
$$prTop\ e = \textbf{let}\ (pTop, \_, \_) = prVersions\ e$$
$$\textbf{in}\ pTop$$

type *ThreeVersions* $= (String, String, String)$

$$paren :: String \to String$$
$$paren\ s = "(" +\!\!+ s +\!\!+ ")"$$

$$prVersions :: E \to ThreeVersions$$
$$prVersions = foldE\ prVerAdd\ prVerMul\ prVerCon$$

The main work is (as usual) done in the "semantic functions":

$$prVerAdd :: ThreeVersions \to ThreeVersions \to ThreeVersions$$
$$prVerAdd\ (\_xTop, xInA, \_xInM)\ (\_yTop, yInA, \_yInM) =$$
$$\quad \textbf{let}\ s = xInA +\!\!+ "+" +\!\!+ yInA \quad \text{-- use } InA \text{ because we are "in } Add"$$
$$\quad \textbf{in}\ (s, paren\ s, paren\ s) \qquad \text{-- parens needed except at top level}$$

$$prVerMul :: ThreeVersions \to ThreeVersions \to ThreeVersions$$
$$prVerMul\ (\_xTop, \_xInA, xInM)\ (\_yTop, \_yInA, yInM) =$$
$$\quad \textbf{let}\ s = xInM +\!\!+ "*" +\!\!+ yInM \quad \text{-- use } InM \text{ because we are "in } Mul"$$
$$\quad \textbf{in}\ (s, s, paren\ s) \qquad\qquad \text{-- parens only needed inside } Mul$$

$$prVerCon :: Integer \to ThreeVersions$$
$$prVerCon\ i \mid i < 0 \quad\ = (s, ps, ps) \quad \text{-- parens needed except at top level}$$
$$\qquad\quad\ \mid otherwise = (s, s,\ \ s) \quad \text{-- parens never needed}$$
$$\quad \textbf{where}\ s = show\ i; ps = paren\ s$$

Exercise 4.12. Another way to make this example go through is to refine the semantic domain from *String* to *Precedence* $\to$ *String*. This can be seen as another variant of the result after the tupling transform: if *Precedence* is an n-element type then *Precedence* $\to$ *String* can be seen as an n-tuple. In our case a three-element *Precedence* would be enough.

4.5   Initial and Free Structures

In Section 4.4 we started with a data-type, and derived an algebraic structure (more precisely an algebra) from it. But we can go in the other direction: start with an algebra and derive a datatype which captures the structure of the algebra, but nothing more. This representation is called the initial algebra.

The Initial Monoid   As a first example, consider an initial algebra for monoids (an initial monoid for short).

We know that we have at least one element: the *unit*. But we can also construct more elements using *op*: *unit 'op' unit*, *unit 'op' (unit 'op' unit)*, etc. So a draft for the initial monoid could be:

> **data** *M = Unit | Op M M*

But we also have the unit laws, which in particular tell us that *unit 'op' unit ==
unit*. So, in fact, we are left with a single element: the *unit*. A representation of the initial monoid is then simply:

> **data** *M = Unit*

As one might guess, there are not many interesting applications of the initial monoid, so let us consider another structure.

The Initial Ring   Gathering all function in various type classes, we find that a *Ring* corresponds to the following algebra — again we start by ignoring laws:

$$zero :: a; \quad (+) :: a \to a \to a; \quad negate :: a \to a$$
$$one \; :: a; \quad (*) \;\; :: a \to a \to a$$

In this case, we can start with *zero* and *one*. As before, using addition on *zero* or multiplication on *one* would yield no more elements. But we can use addition on *one*, and get *one + one*, *one + one + one*, etc. Because of associativity, we don't have to — and ought not to — write parentheses. Let's write an addition of *n* ones as *n*. What about multiplying? Are we going to get more kinds of numbers from that? No, because of distributivity. For example:

$$(one + one) * (one + one) == one + one + one + one$$

So far we have only the natural numbers, but the last operation, *negate*, adds the negative numbers as well. By following this line of reasoning to its conclusion, we will find that the initial *Ring* is the set of integers.

4.5.1 A general initial structure

In Haskell, the type $C\ a \Rightarrow a$ is a generic way to represent the initial algebra for a class C. To get a more concrete feeling for this, let us return to *IntExp*, and consider a few values of type *IntExp* $a \Rightarrow a$.

> *seven* :: *IntExp* $a \Rightarrow a$; *seven* = *add* (*con* 3) (*con* 4)
>
> *testI* :: *Integer*;   *testI* = *seven*
> *testE* :: *E*;   *testE* = *seven*
> *testP* :: *String*;   *testP* = *seven*
>
> *check* :: *Bool*
> *check* = *and* [*testI* == 7
>   , *testE* == *Add* (*Con* 3) (*Con* 4)
>   , *testP* == "3+4"
>   ]

By defining a class *IntExp* (and some instances) we can use the methods (*add*, *mul*, *con*) of the class as "generic constructors" which adapt to the context. An overloaded expression, like *seven* :: *IntExp* $a \Rightarrow a$, which only uses these generic constructors can be instantiated to different types, ranging from the syntax tree type *E* to any possible semantic interpretations (like *Integer*, *String*, etc.). In general, for any given value x of type *IntExp* $a \Rightarrow a$, all the variants of x instantiated at different types are guaranteed to be related by homomorphisms, because one simply replaces *add*, *mul*, *con* by valid instances.

The same kind of reasoning justifies the overloading of Haskell integer literals. They can be given the type *Ring* $a \Rightarrow a$, and doing it in a mathematically meaningful way, because *Ring* $a \Rightarrow a$ is the initial algebra for *Ring*.

4.5.2 *Free Structures

Another useful way of constructing types is through "free structures". They are similar to initial structures, but they also allows one to embed an arbitrary set of *generators G*. That is, it is as if we would throw an additional *generate* function in the algebra:

> **class** *Generate a* **where**
>   *generate* :: $G \rightarrow a$

We could parameterise the class over an abstract generator set g, but will refrain from doing so to avoid needless complications.

Free Monoid   As an example, consider the free monoid. Our algebra has the following signature:

$$generate :: G \rightarrow a$$
$$unit \quad :: a$$
$$op \quad \quad :: a \rightarrow a \rightarrow a$$

As a first version, we can convert each function to a constructor and obtain the following type:

data *FreeMonoid g* = *Unit*
       | *Op (FreeMonoid g) (FreeMonoid g)*
       | *Generator g* **deriving** *Show*
instance *Monoid (FreeMonoid g)* **where** *unit* = *Unit*; *op* = *Op*

Let us consider a fold for *FreeMonoid*. We can write its type as follows:

$$evalM :: (Monoid\ a, Generate\ a) \Rightarrow (FreeMonoid\ G \rightarrow a)$$

but we can also drop the *Generate* constraint and take the *generate* method as an explicit argument:

$$evalM :: Monoid\ a \Rightarrow (G \rightarrow a) \rightarrow (FreeMonoid\ G \rightarrow a)$$

This form is similar to the evaluators of expressions with variables of type *G*, which we have seen for example in Section 1.7.3. Once given a function $f :: G \rightarrow a$ (which we call an "assignment function"), the homomorphism condition forces *evalM* to be a fold:

$$evalM \_ Unit \quad \quad = unit$$
$$evalM\ f\ (Op\ e_1\ e_2) \quad = op\ (evalM\ f\ e_1)\ (evalM\ f\ e_2)$$
$$evalM\ f\ (Generator\ x) = f\ x$$

However, before being completely satisfied, we must note that the *FreeMonoid* representation is ignoring monoid laws. By following the same kind of reasoning as before, we find that we have in fact only two distinct forms for the elements of the free monoid:

- *unit*

- *generate x_1 'op' generate x_2 'op' . . . 'op' generate x_n*

Because of associativity we have no parentheses in the second form; and because of the unit laws we need not have *unit* composed with *op* either.

Thus, the free monoid over a generator set G can be represented by a list of G.

We seemingly also ignored the laws when defining *evalM*. Is this a problem? For example, is it possible that e_1 '*Op*' $(e_2$ '*Op*' $e_3)$ and $(e_1$ '*Op*' $e_2)$ '*Op*' e_3 which are by monoid laws equal, map to different values? By definition of *evalM*, the condition reduces to checking *evalM* f e_1 '*op*' (*evalM* f e_2 '*op*' *evalM* f e_3) == (*evalM* f e_1 '*op*' *evalM* f e_2) '*op*' *evalM* f e_3. But then, this turns out to be satisfied if *op* is associative. In sum, *evalM* will be correct if the target *Monoid* instance satisfies the laws. This is true in general: folds are always homomorphisms even if the datatype representation that they work on ignore laws.

Functions of one variable as free algebras   Earlier we have used (many variants of) data types for arithmetic expressions. Using the *free* construction, we can easily conceive a suitable type for any such expression language. For example, the type for arithmetic expressions with $(+), (-), (*)$ and variables is the free *Ring* with the set of variables as generator set.

Let us consider again our deep embedding for expressions of one variable from Section 1.7.1. According to our analysis, it should be a free structure, and because we have only one variable, we can take the generator set (G) to be the unit type.

> **type** $G = ()$
> **instance** *Generate FunExp* **where** *generate* $() = X$

We can easily show that *FunExp* is *Additive* and *Multiplicative*:

> **instance** *Additive*   *FunExp* **where** $(+) = (:+:); zero = Const\ 0$
> **instance** *Multiplicative FunExp* **where** $(*) = (:*:);\ one = Const\ 1$

Exercise 4.13. Implement *FunExp* instances for *AddGroup*, and (possibly extending the datatype) for *MulGroup* and *Transcendental*.

We can then define a compositional evaluator. It would start as follows:

> *eval* $(e_1 :*: e_2) = eval\ e_1 * eval\ e_2$
> *eval* $(e_1 :+: e_2) = eval\ e_1 + eval\ e_2$

Remark: to translate the *Const* :: $\mathbb{R} \to FunExp$ constructor we need a way to map any $\mathbb{R}$ to the above structures. Here we will restrict ourselves to integers.

The most general type of evaluator will give us:[2]

> **type** *OneVarExp a* = (*Generate a*, *Ring a*)
> *eval* :: *FunExp* → (*OneVarExp a* ⇒ *a*)

With this class in place we can define generic expressions using generic constructors just like in the case of *IntExp* above. For example, we can define

> *varX* :: *OneVarExp a* ⇒ *a*
> *varX* = *generate* ()
> *twoX* :: *OneVarExp a* ⇒ *a*
> *twoX* = *two* ∗ *varX*

and instantiate *twoexp* to either syntax or semantics:

> **type** *Func* = $\mathbb{R} \to \mathbb{R}$
> *testFE* :: *FunExp*; *testFE* = *twoX*
> *testFu* :: *Func*;   *testFu* = *twoX*

provided a suitable instance for *Generate Func*:

> **instance** *Generate Func* **where**
>   *generate* () = *id*

As before, we can always define a homomorphism from *FunExp* to *any* instance of *OneVarExp*, in a unique way, using the fold pattern.

This is because the datatype *FunExp* is an initial *OneVarExp*. Working with *OneVarExp a* ⇒ *a* can be more economical than using *FunExp*: one does not need any explicit *eval* function.

We now have two DSLs which capture the similar concepts. One of them is given by the data type *FunExp*. The other one is given by the type class (synonym) *OneVarExp*. In fact, the instances and the evaluator would form an isomorphism between *FunExp* (the version restricted to integer constants) and *OneVarExp a* ⇒ *a*.

The difference is that the first one builds a syntax tree, while the other one refers to the semantics (algebraic) value. For example, (:+:) *stands for* a function, while (+) *is* that function.

4.5.3   *A generic Free construction

We can use the same trick as for initial algebras to construct free algebras: (*C a*, *Generate a*) ⇒ *a* is the free *C*-structure. However, it is often more convenient to pass the embedding function explicitly rather than via the *Generate*

---

[2]We call this constraint "OneVarExp" because we have fixed *G* = (). In general the number of variables is the cardinality of *G*.

class. In this case, we obtain the type: $C\ a \Rightarrow (g \to a) \to a$ if g is the set of generators. In modern versions of Haskell, we can even parameterise over the C class, and write:

newtype *Free c g* = *Free* (*forall a. c a* $\Rightarrow$ ($g \to a$) $\to a$)

Embedding a generator is then done as follows:

embed :: $g \to$ *Free c g*
embed g = *Free* (λ*generate* $\to$ *generate g*)

Unfortunately the *Free c* type is not automatically an instance of *c*: we have to implement those manually. Let us see how this plays out for monoid:

instance *Monoid* (*Free Monoid g*) **where**
  unit = *Free* ($\backslash\_ \to$ *unit*)
  Free f '*op*' *Free g* = *Free* ($\lambda x \to f\ x$ '*op*' *g x*)

We can also check the monoid laws for the free monoid. For example, here is the proof that the right identity law holds:

Free f '*op*' *unit*   == {- def. -}
Free f '*op*' *Free* ($\backslash\_ \to$ *unit*) == {- def. -}
Free ($\lambda x \to f\ x$ '*op*' *unit*)   == {- law of the underlying monoid -}
Free ($\lambda x \to f\ x$)   == {- eta-reduction -}
Free f

Exercise 4.14. Prove group laws for *Free AdditiveGroup*.

We can also recover the whole structure which was used to build an element of this type, for example we could use lists (recall that they are isomorphic to free monoids):

extract :: *Free Monoid g* $\to$ [g]
extract (*Free f*) = *f* ($\lambda g \to$ [g])

As an example, we can *extract* the value of the following example:

example :: *Free Monoid Int*
example = *embed* 1 '*op*' *embed* 10 '*op*' *unit* '*op*' *embed* 11

  -- $\ggg$ *extract example*
  -- [$1, 10, 11$]

Exercise 4.15. Show that *Free Ring* () covers most of the type *FunExp* from Section 1.7.1.

4.6   Computing derivatives, reprise

As discussed in Section 4.5.2, it can sometimes be good to use the representation *OneVarExp a* $\Rightarrow$ *a* rather than the *FunExp* data type. However, in Section 3.6 we argued that the rules for derivatives were naturally operating on a syntactic representation.

The question is: can we implement *derive* in the shallow embedding? As a reminder, the reason that the shallow embedding ($\mathbb{R} \rightarrow \mathbb{R}$) works is that the *eval* function is a *fold*: first evaluate the sub-expressions of *e*, then put the evaluations together without reference to the sub-expressions.

Let us now check whether the semantics of derivatives is compositional. This evaluation function for derivatives is given by composition as below:

type *Func* $= \mathbb{R} \rightarrow \mathbb{R}$
eval' :: *FunExp* $\rightarrow$ *Func*
eval' $=$ *eval* $\circ$ *derive*

In a diagram:

$$
\begin{array}{ccc}
FunExp & \xrightarrow{\;eval\;} & (\mathbb{R} \rightarrow \mathbb{R}) \\
\Big\downarrow{\scriptstyle derive} & \;\;\searrow{\scriptstyle eval'} & \Big\downarrow{\scriptstyle D} \\
FunExp & \xrightarrow{\;eval\;} & (\mathbb{R} \rightarrow \mathbb{R})
\end{array}
$$

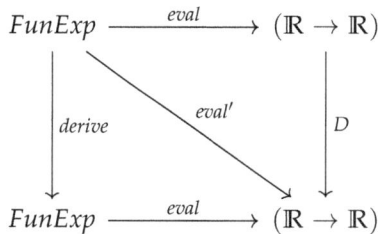

The diagram shows types as nodes and functions between those types as labelled edges. On the left side we have the syntactic part: the *FunExp* type and the *derive* function. On the right side we have the semantic domain: the type of functions from $\mathbb{R}$ to $\mathbb{R}$, and the mathematical *D* function. The diagonal is the function we would like to implement compositionally: *eval'*. The diagonal is specified by composing the functions on the "left-bottom" path. Let us consider the *Exp* case (the *eval'*$_{Exp}$-lemma):

eval' (*Exp e*)	$=$ {- def. *eval'*, function composition -}
eval (*derive* (*Exp e*))	$=$ {- def. *derive* for *Exp* -}
eval (*Exp e* :*: *derive e*)	$=$ {- def. *eval* for :*: -}
eval (*Exp e*) $*$ *eval* (*derive e*)	$=$ {- def. *eval* for *Exp* -}
exp (*eval e*) $*$ *eval* (*derive e*)	$=$ {- def. *eval'* -}
exp (*eval e*) $*$ *eval' e*	$=$ {- let $f =$ *eval e*, $f' =$ *eval' e* -}
exp f $* f'$	

Thus given *only* the derivative $f' = eval'\ e$, it looks hard to compute $exp\ f * f'$. More concretely, if we take $e_1 = X$ and $e_2 = X$:+: *Const* 1, then $eval'\ e_1 ==$ *const* 1 == $eval'\ e_2$ but $eval'\ (Exp\ e_1)\ 0 == 1 \neq e == eval'\ (Exp\ e_2)$. Thus, it is impossible to compute $eval'$ compositionally.

Another example of the problem is *derive* $(f$:*: $g)$ where the result involves not only *derive f* and *derive g*, but also f and g on their own. In general, the problem is that some of the rules for computing the derivative depend not only on the derivative of the subexpressions, but also on the subexpressions before taking the derivative.

Consequently, $eval'$ is in fact non-compositional (just like *isPrime*). There is no way to implement $eval' :: FunExp \to Func$ as a fold *if Func is the target type*. One way of expressing this fact is to say that in order to implement $eval' :: FunExp \to Func$ we need to also compute $eval :: FunExp \to Func$. Thus we need to implement a pair of *eval*-functions $(eval, eval')$ together.

In practice, the solution is to extend the return type of $eval'$ from one semantic value f of type $Func = \mathbb{R} \to \mathbb{R}$ to two such values $(f, f') :: (Func, Func)$ where $f' = D\ f$. That is, we are using the "tupling transform": we are computing just one function $evalD :: FunExp \to (Func, Func)$ returning a pair of f and $D\ f$ at once. (At this point, you are advised to look up and solve Exercise 1.6 in case you have not done so already.)

$$\textbf{type}\ FD\ a = (a \to a, a \to a)$$
$$evalD :: FunExp \to FD\ \mathbb{R}$$
$$evalD\quad e\qquad = (eval\ e, eval'\ e)$$

Is $evalD$ compositional? We compute, for example:

$$
\begin{array}{ll}
evalD\ (Exp\ e) & = \{\text{- specification of } evalD \text{ -}\} \\
(eval\ (Exp\ e), eval'\ (Exp\ e)) & = \{\text{- def. } eval \text{ for } Exp, eval'_{Exp}\text{-lemma -}\} \\
(exp\ (eval\ e), exp\ (eval\ e) * eval'\ e) & = \{\text{- introduce local names -}\} \\
\textbf{let}\ f\ = eval\ e & \\
\quad f' = eval'\ e & \\
\textbf{in}\ (exp\ f, exp\ f * f') & = \{\text{- def. } evalD \text{ -}\} \\
\textbf{let}\ (f, f') = evalD\ e & \\
\textbf{in}\ (exp\ f, exp\ f * f') &
\end{array}
$$

This semantics *is* compositional and the *Exp* case is as follows:

$$evalD_{Exp} :: FD\ \mathbb{R} \to FD\ \mathbb{R}$$
$$evalD_{Exp}\quad (f, f') = (exp\ f, exp\ f * f')$$

In general, while *eval'* is non-compositional, *evalD* is a more complex, but compositional, semantics. We can then get *eval'* back as the second component of *evalD e*:

$$eval' :: FunExp \rightarrow Func$$
$$eval' = snd \circ evalD$$

Because all compositional functions can be expressed as a fold for a given algebra, we can now define a shallow embedding for the combined computation of functions and derivatives, using the numerical type classes.

> **instance** *Additive a* $\Rightarrow$ *Additive* (*FD a*)   **where**
>   *zero* $=$ *zeroFD*; $(+) = addFD$
> **instance** (*Additive a*, *Multiplicative a*) $\Rightarrow$ *Multiplicative* (*FD a*) **where**
>   *one* $=$ *oneFD*; $(*) = mulFD$
>
> *zeroFD* :: *Additive a* $\Rightarrow$ *FD a*
> *oneFD* :: (*Additive a*, *Multiplicative a*) $\Rightarrow$ *FD a*
> *zeroFD* $=$ (*const zero*, *const zero*)
> *oneFD* $=$ (*const one*, *const zero*)
>
> *addFD* :: *Additive a* $\Rightarrow$ *Dup a* $\rightarrow$ *Dup a* $\rightarrow$ *Dup a*
> *mulFD* :: (*Additive a*, *Multiplicative a*) $\Rightarrow$ *Dup a* $\rightarrow$ *Dup a* $\rightarrow$ *Dup a*
> *addFD* (f, f') (g, g') $= (f + g, f'\quad + \quad g')$
> *mulFD* (f, f') (g, g') $= (f * g, f' * g + f * g')$

Exercise 4.16. Implement the rest of the numeric instances for *FD a*.

4.6.1   Automatic differentiation

The simultaneous computation of values and derivatives is an important technique called "automatic differentiation". Automatic differentiation has grown in importance with the rise of machine learning, which often uses derivatives (or gradients) to find a values of parameter which minimises a user-defined objective function. However, in such systems, one is often not interested in computing whole functions and their derivatives (as we have done so far), but rather a function at a point (say $f\ x_0$) and the derivative at the same point (say $D f\ x_0$).

The question then arises: is it enough to *only* compute the pair $(f\ x_0, D f\ x_0)$? In other words, is automatic differentiation compositional? To answer this question, we must find yet again if there is a homomorphism between whole functions and their value at a point.

Fortunately, we have already seen part of the answer in Exercise 4.6. Namely, the homomorphism is *apply c*, with the definition:

$$apply :: a \rightarrow (a \rightarrow b) \rightarrow b$$
$$apply\ a = \lambda f \rightarrow f\ a$$

Because *apply c* is so simple, it is an homomorphism not only for *Additive*, but also *Ring* (and any numeric class we have seen so far). We already took advantage of this simple structure to define homomorphism in the other direction in Section 3.5.3, where we defined a *Ring* instance for functions with a *Ring* codomain.

Can we do something similar for *FD*? The elements of *FD a* are pairs of functions, so we can take

type *Dup a* $= (a, a)$
type *FD a* $= (a \rightarrow a, a \rightarrow a)$
applyFD $:: a \rightarrow FD\ a \rightarrow Dup\ a$
applyFD c $(f, f') = (f\ c, f'\ c)$

We now have the domain of the homomorphism (*FD a*) and the homomorphism itself (*applyFD c*), but we are missing the structure on the codomain, which now consists of pairs *Dup a* $= (a, a)$. In fact, we can *compute* this structure from the homomorphism condition. For example:

$$
\begin{array}{ll}
h\ ((f, f') * (g, g')) & = \{\text{- def. } (*)\text{ for } FD\ a \text{ -}\} \\
h\ (f * g, f' * g + f * g') & = \{\text{- def. } h = applyFD\ c \text{ -}\} \\
((f * g)\ c, (f' * g + f * g')\ c) & = \{\text{- def. } (*)\text{ and } (+)\text{ for functions -}\} \\
(f\ c * g\ c, f'\ c * g\ c + f\ c * g'\ c) & = \{\text{- let } x = f\ c; y = g\ c; x' = f'\ c; y' = g'\ c \text{ -}\} \\
(x * y, x' * y + x * y') & = \{\text{- \textbf{introduce} } \circledast\textbf{: see def. below} \text{ -}\} \\
(x, x') \circledast (y, y') & = \{\text{- expand shorter names again -}\} \\
(f\ c, f'\ c) \circledast (g\ c, g'\ c) & = \{\text{- def. } h = applyFD\ c \text{ -}\} \\
h\ (f, f') \circledast h\ (g, g')
\end{array}
$$

The identity will hold if we take

$$(\circledast) :: Ring\ a \Rightarrow Dup\ a \rightarrow Dup\ a \rightarrow Dup\ a$$
$$(x, x') \circledast (y, y') = (x * y, x' * y + x * y')$$

Thus, if we define a "multiplication" on pairs of values using $(\circledast)$, we get that (*applyFD c*) is a homomorphism from *FD a* to *Dup a* for all *c*. To make it a *Multiplicative*-homomorphism we just need to calculate a definition for *oneDup* to make it satisfy to homomorphism law:

$$oneDup \qquad\qquad = \{\text{-} H_0(applyFD\ c, oneFD, oneDup)\ \text{-}\}$$
$$applyFD\ c\ oneFD \qquad = \{\text{-} \text{Def. of } oneFD \text{ and } applyFD \text{ -}\}$$
$$(const\ one\ c, const\ zero\ c) = \{\text{-} \text{Def. of } const \text{ -}\}$$
$$(one, zero)$$

We can now define an instance

$$oneDup :: Ring\ a \Rightarrow Dup\ a$$
$$oneDup = (one, zero)$$
instance $Ring\ a \Rightarrow Multiplicative\ (Dup\ a)$ **where**
  $one = oneDup$
  $(*) = (\circledast)$

Exercise 4.17. Complete instance declarations for *Dup* $\mathbb{R}$: *Additive, AddGroup,* etc.

In sum, because this computation goes through also for the other cases we can actually work with just pairs of values (at an implicit point $c :: a$) instead of pairs of functions. Thus we can define a variant of *FD* a to be **type** *Dup* $a = (a, a)$.

4.7 Summary

In this chapter we have compared and contrasted a number of mathematical concepts and their computer science representations or alternative interpretations. Mathematical structures can often be used to capture the core of a DSL, initial algebras can be used (with **data**-types) for abstract syntax (deep embeddings) but can also be constructed with type classes (*Class* $a \Rightarrow a$) without reference to concrete **data**. Other algebras capture different shallow embeddings, and semantics (the semantic function *eval*) is normally a homomorphism from the initial algebra.

4.7.1 Homomorphism as roadmaps

Homomorphisms are key to describe mathematical structures, specify programs, and derive of correct programs. The relation $h : S_1 \rightarrow S_2$ (standing for "h is a homomorphism from S_1 to S_2"), can be used in many ways, depending on what is known and what is unknown.

- $? : S_1 \rightarrow S_2$. Given two structures S_1 and S_2, can we derive some function which is a homomorphism between those two structures? We asked such a question in Exercise 4.6 (*apply c : Additive* $(x \rightarrow a) \rightarrow$ *Additive a*) and Section 4.2.2 (exponentials).

- $h : S_1 \rightarrow ?$. What is a structure S_2 compatible with a given structure S_1 and given homomorphism h? (e.g., we derived the operations on *Dup a* $= (a, a)$ from *applyFD c : FD a* $\rightarrow$ *Dup a* and operations on *FD a* in Section 4.6.1.)

- $? : S_1 \rightarrow ?$. Can we find a good structure on S_2 so that it becomes homomorphic with S_1? This is how we found the structure *FD* in *evalD* : *FunExp* $\rightarrow$ *FD a*.

- $h : ? \rightarrow S_2$. Given h and S_2, can we find a structure S_1 compatible with a given homomorphism h? We will encounter an example in Chapter 5 (evaluation function for polynomials).

4.7.2 Structures and representations

One take home message of this chapter is that one should, as a rule, start with structural definitions first, and consider representation second. For example, in Section 4.6 we defined a *Ring* structure on pairs $(\mathbb{R}, \mathbb{R})$ by requiring the operations to be compatible with the interpretation $(f\ a, f'\ a)$. This requirement yields the following definition for multiplication for pairs:

$$(x, x') \circledast (y, y') = (x * y, x' * y + x * y')$$

But there is nothing in the "nature" of pairs of $\mathbb{R}$ that forces this definition upon us. We chose it, because of the intended interpretation.

This multiplication is not the one we need for *complex numbers*. It would be instead:

$$(x, x') *. (y, y') = (x * y - x' * y', x * y' + x' * y)$$

Again, there is nothing in the nature of pairs that foists this operation on us. In particular, it is, strictly speaking, incorrect to say that a complex number *is* a pair of real numbers. The correct interpretation is that a complex number can be *represented* by a pair of real numbers, provided that we define the operations on these pairs in a suitable way.

The distinction between definition and representation is similar to the one between specification and implementation, and, in a certain sense, to the one between syntax and semantics. All these distinctions are frequently obscured, for example, because of prototyping (working with representations / implementations / concrete objects in order to find out what definition / specification / syntax is most adequate). They can also be context-dependent (one man's specification is another man's implementation). Insisting on the difference between definition and representation can also appear quite pedantic (as in the discussion of complex numbers in Section 1.4). In general though, it is a good idea to be aware of these distinctions, even if they are suppressed for reasons of brevity or style. We will encounter this distinction again in Section 5.1.

4.8 Beyond Algebras: Co-algebra and the Stream calculus

In the coming chapters there will be quite a bit of material on infinite structures. These are often captured not by algebras, but by co-algebras. We will not build up a general theory of co-algebras in this book, but because we will be using infinite streams in the upcoming chapters we will expose right here their co-algebraic structure.

Streams as an abstract datatype   Consider the API for streams of values of type A represented by some abstract type X:

```
data X
data A
head :: X → A
tail  :: X → X
cons :: A → X → X
law1  s =      s == cons (head s) (tail s)
law2 a s =     s == tail  (cons a s)
law3 a s =     a == head (cons a s)
```

With this API we can use *head* to extract the first element of the stream, and *tail* to extract the rest as a new stream of type X. Using *head* and *tail* recursively we can extract an infinite list of values of type A:

```
toList :: X → [A]
toList x = head x : toList (tail x)
```

In the other direction, if we want to build a stream we only have one constructor: *cons* but no "base case". In Haskell, thanks to laziness, we can still define streams directly using *cons* and recursion As an example, we can construct a constant stream as follows:

> $constS :: A \rightarrow X$
> $constS\ a = ca$
>   **where** $ca = cons\ a\ ca$

Instead of specifying a stream in terms of how to construct it, we could describe it in terms of how to take it apart; by specifying its *head* and *tail*. In the constant stream example we would get something like:

> $head\ (constS\ a) = a$
> $tail\ \ \ (constS\ a) = constS\ a$

but this syntax is not supported in Haskell.

The last part of the API are a few laws we expect to hold. The first law simply states that if we first take a stream s apart into its head and its tail, we can get back to the original stream by *cons*ing them back together. The second and third are variant on this theme, and together the three laws specify how the three operations interact.

4.9   A solved exercise

We have seen three different ways to use a generic $f :: Transcendental\ a \Rightarrow a \rightarrow a$ to compute the derivative at some point (say, at 2.0, $f'\ 2$):

1. fully symbolic (using *FunExp*),

2. using pairs of functions ($FD\ a = (a \rightarrow a, a \rightarrow a)$),

3. or just pairs of values ($Dup\ a = (a, a)$).

Given the following definition of f, compute $f'\ 2$.

> $f :: Transcendental\ a \Rightarrow a \rightarrow a$
> $f\ x = sin\ x + two * x$

(So, we have: $f\ 0 = 0, f\ 2 = 4.909297426825682$, etc.)

Solution:

1. Using *FunExp*

   Recall expressions (or functions) of one variables, from Section 1.7.1:

   > **data** *FunExp* = *Const Rational*
   > | *X*
   > | *FunExp* :+: *FunExp*
   > | *FunExp* :∗: *FunExp*
   > | *FunExp* :/: *FunExp*
   > | *Exp FunExp* | *Sin FunExp* | *Cos FunExp*
   >   -- and so on
   >   **deriving** (*Eq, Show*)

   What is the expression *e* for which *f* = *eval e*?

   We have

   $$eval\ e\ x = f\ x$$
   $$\Leftrightarrow eval\ e\ x = sin\ x + 2 * x$$
   $$\Leftrightarrow eval\ e\ x = eval\ (Sin\ X)\ x + eval\ (Const\ 2\ :*:\ X)\ x$$
   $$\Leftrightarrow eval\ e\ x = eval\ ((Sin\ X)\ :+:\ (Const\ 2\ :*:\ X))\ x$$
   $$\Leftarrow e = Sin\ X\ :+:\ (Const\ 2\ :*:\ X)$$

   Finally, we can apply *derive* :: *FunExp* → *FunExp*, defined in Section 3.6, and obtain

   $$e = Sin\ X\ :+:\ (Const\ 2\ :*:\ X)$$
   $$f'\ 2 = eval\ (derive\ e)\ 2$$

   This can hardly be called "automatic", look at all the work we did in deducing *e*![3]

   However, consider this definition:

   $$fe :: FunExp$$
   $$fe = f\ X$$

   As *X* :: *FunExp*, the Haskell interpreter will look for *FunExp* instances of *Additive* and other numeric classes and build the syntax tree for *f* instead of computing its semantic value.

   In general, to find the derivative of a function *f* :: *Transcendental a* ⇒ *a* → *a*, we can use

   $$drv\ f = eval\ (derive\ (f\ X))$$

---

[3]Besides, manipulating symbolic representations (even in a program), is not was is usually called automatic differentiation.

2. Using *FD* (pairs of functions)

   Recall

   $$\textbf{type } FD \ a = (a \rightarrow a, a \rightarrow a)$$
   $$applyFD \ c \ (f,g) = (f \ c, g \ c)$$

   The operations (the numeric type class instances) on *FD a* are such that, if *eval e* $= f$, then

   $$(eval \ e, eval' \ e) = (f, f')$$

   We are looking for (g, g') such that

   $$f \ (g, g') = (f, f') \quad -- (*)$$

   so we can then do

   $$f' \ 2 = snd \ (applyFD \ 2 \ (f \ (g, g')))$$

   We can fulfil (*) if we can find a pair (g, g') that is a sort of "unit" for *FD a*:

   $$sin \ (g, g') = (sin, \ cos)$$
   $$exp \ (g, g') = (exp, exp)$$

   and so on.

   In general, the chain rule gives us

   $$f \ (g, g') = (f \circ g, (f' \circ g) * g')$$

   Therefore, if we look at the first components of two examples we see that we need to satisfy $sin \circ g$ == sin and $exp \circ g$ == exp. For this we can take $g = id$. Then, looking at the second components we need to satisfy $sin' * g'$ == cos and $exp' * g'$ == exp which simplifies to $cos * g'$ == cos and $exp * g'$ == exp which is solved by $g' = const \ 1$.

   Finally

   $$f' \ 2 = snd \ (applyFD \ 2 \ (f \ (id, const \ 1)))$$

   In general

   $$drvFD \ f \ x = snd \ (applyFD \ x \ (f \ (id, const \ 1)))$$

   computes the derivative of f at x.

3. Using *Dup* (pairs of values).

We have **instance** *Transcendental a* $\Rightarrow$ *Transcendental* (a, a), moreover, the instance declaration looks exactly the same as that for *FD a*:

> **instance** *Transcendental a* $\Rightarrow$ *Transcendental* $(FD\ a)$ **where**
> $exp\ (f, f') = (exp\ f, (exp\ f) * f')$ -- pairs of functions
> $sin\ (f, f') = (sin\ f, (cos\ f) * f')$
>
> -- ...
>
> **instance** *Transcendental a* $\Rightarrow$ *Transcendental* (a, a) **where**
> $exp\ (x, x') = (exp\ x, (exp\ x) * x')$ -- just pairs
> $sin\ (x, x') = (sin\ x, (cos\ x) * x')$
>
> -- ...

In fact, the latter instance (just pairs) a generalisation of the former instance (FD). To see this, recall that $FD\ a = (a \to a, a \to a)$, and note that if we have a *Transcendental* instance for some A, we get a *Transcendental* instance for $x \to A$ for all x from Fig. 3.1. Then from the instance for pairs we get an instance for any type of the form $(x \to A, x \to A)$. As a special case when $x = A$ this includes all $(A \to A, A \to A)$ which is *FD A*. Thus it is enough to have the instance *Transcendental* $(x \to A)$ and the pair instance to get the "pairs of functions" instance (and more).

The pair instance is also the "maximally general" such generalisation.

Still, we need to use this machinery to finally compute $f'\ 2$. We are now looking for a pair of values (g, g') such that

$$f\ (g, g') = (f\ 2, f'\ 2)$$

In general we have the chain rule:

$$f\ (g, g') = (f\ g, (f'\ g) * g')$$

Therefore

$$
\begin{aligned}
&f\ (g, g') = (f\ 2, f'\ 2) \\
\Leftrightarrow\ &(f\ g, (f'\ g) * g') = (f\ 2, f'\ 2) \\
\Leftarrow\ &g = 2, g' = 1
\end{aligned}
$$

Introducing *var x* $= (x, one)$ we can, as in the case of *FD*, simplify matters a little:

$$f'\ x = snd\ (f\ (var\ x))$$

In general

$$drvP\ f\ x = snd\ (f\ (var\ x))$$

computes the derivative of f at x.

Numeric instances for *Dup* For reference: the rest of the instance declarations for *Dup* (the *Multiplicative* instance was provided above):

> **instance** *Additive a* $\Rightarrow$ *Additive* (*Dup a*) **where**
>   *zero* = *zeroDup*; (+) = *addDup*
>
> *zeroDup* :: *Additive a* $\Rightarrow$ *Dup a*
> *zeroDup* = (*zero*, *zero*)
>
> *addDup* :: *Additive a* $\Rightarrow$ *Dup a* $\rightarrow$ *Dup a* $\rightarrow$ *Dup a*
> *addDup* (x, x') (y, y') = $(x + y, x' + y')$
>
> **instance** *AddGroup a* $\Rightarrow$ *AddGroup* (*Dup a*) **where**
>   *negate* = *negateDup*
>
> *negateDup* :: *AddGroup a* $\Rightarrow$ *Dup a* $\rightarrow$ *Dup a*
> *negateDup* (x, x') = (*negate x*, *negate* x')
>
> **instance** (*AddGroup a*, *MulGroup a*) $\Rightarrow$ *MulGroup* (*Dup a*) **where**
>   *recip* = *recipDup*
>
> *recipDup* :: (*AddGroup a*, *MulGroup a*) $\Rightarrow$ *Dup a* $\rightarrow$ *Dup a*
> *recipDup* (x, x') = (y, y')
>   **where** y = *recip x*
> y' = *negate* $(y * y) * x'$
>
> **instance** *Transcendental a* $\Rightarrow$ *Transcendental* (*Dup a*) **where**
> π = *piDup*; *sin* = *sinDup*; *cos* = *cosDup*; *exp* = *expDup*
>
> *piDup* :: *Transcendental a* $\Rightarrow$ *Dup a*
> *piDup* = $(\pi, zero)$
>
> *sinDup*, *cosDup*, *expDup* :: *Transcendental a* $\Rightarrow$ *Dup a* $\rightarrow$ *Dup a*
> *sinDup* (x, x') = (*sin x*,   *cos x* $* x'$)
> *cosDup* (x, x') = (*cos x*, *negate* (*sin x*) $* x'$)
> *expDup* (x, x') = (*exp x*,   *exp x* $* x'$)
>
> *var* :: *Multiplicative a* $\Rightarrow$ *a* $\rightarrow$ *Dup a*
> *var x* = (x, one)

4.10   Exercises

Exercise 4.18. Homomorphisms. Consider the following definitions:

> -- $h : A \to B$ is a homomorphism from $Op : A \to A \to A$ to $op : B \to B \to B$
> $H_2(h, Op, op) = \forall x. \; \forall y. \; h \; (Op \; x \; y) \; \texttt{==} \; op \; (h \; x) \; (h \; y)$
> -- $h : A \to B$ is a homomorphism from $F : A \to A$ to $f : B \to B$
> $H_1(h, F, f) \quad = \forall x. \; h \; (F \; x) \; \texttt{==} \; f \; (h \; x)$
> -- $h : A \to B$ is a homomorphism from $E : A$ to $e : B$
> $H_0(h, E, e) \quad = h \; E \; \texttt{==} \; e$

Prove or disprove the following claims:

- $H_2((2*), (+), (+))$

- $H_2((2*), (*), (*))$

- $H_2(exp, (+), (*))$

- $H_2(eval', (:+:), (+))$

- $H_1(\sqrt{\cdot}, (4*), (2*))$

- $\exists f. \; H_1(f, (2*) \circ (1+), (1+) \circ (2*))$

Exercise 4.19. Complete the numeric instance declarations for *FunExp*.

Exercise 4.20. Complete the instance declarations for *Dup* $\mathbb{R}$, deriving them from the homomorphism requirement for *applyFD* (in Section 4.6.1).

Exercise 4.21. We now have three different ways of computing the derivative of a function such as $f \; x = sin \; x + exp \; (exp \; x)$ at a given point, say $x = \pi$.

1. Find $e :: FunExp$ such that $eval \; e = f$ and use *eval'*.

2. Find an expression of type $FD \; \mathbb{R}$ and use *apply*.

3. Apply f directly to the appropriate (x, x') and use *snd*.

Do you get the same result?

Exercise 4.22. In Exercise 1.14 we looked at the datatype $SR \; v$ for the language of semiring expressions. We will now use some of the concepts discussed in this chapter to expand on this language.

1. Define a type class *SemiRing* that corresponds to the semiring structure.

2. Define a *SemiRing* instance for the datatype *SR v* that you defined in exercise 1.3.

3. Find two other instances of the *SemiRing* class.

4. Specialise the evaluator that you defined in Exercise 1.14 to the two *SemiRing* instances defined above. Take three semiring expressions of type *SR String*, give the appropriate assignments and compute the results of evaluating, in each case, the three expressions.

Exercise 4.23. Show that arithmetic modulo n satisfies the semiring laws (it is even a ring). In more details: show that $\mathbb{Z}_n = \{0, 1, \ldots, n-1\}$ with *plus x y* $= (x + y)\%n$ and *times x y* $= (x * y)\%n$ forms a semiring.

With $h\ x = x\%n$, show that h is a homomorphism from $\mathbb{Z}$ to $\mathbb{Z}_n$.

Exercise 4.24. In Exercise 1.15, we looked at a datatype for the language of lattice expressions.

1. Define a type class *Lattice* that corresponds to the lattice structure.

2. Define a *Lattice* instance for the datatype for lattice expressions that you defined in Exercise 1.15.

3. Find two other instances of the *Lattice* class.

4. Specialise the evaluator you defined in Exercise 1.15 to the two *Lattice* instances defined above. Take three lattice expressions, give the appropriate assignments and compute the results of evaluating, in each case, the three expressions.

Exercise 4.25. In Exercise 1.16, we looked at a datatype for the language of abelian monoid expressions. We will now use some of the concepts discussed in this chapter to expand on this language.

1. Define a type class *AbMonoid* that corresponds to the abelian monoid structure.

2. Define an *AbMonoid* instance for the datatype for abelian monoid expressions that you defined in Exercise 1.16.

3. Find one other instance of the *AbMonoid* class and give an example which is **not** an instance of *AbMonoid*.

4. Specialise the evaluator that you defined in Exercise 1.16 to the *AbMonoid* instance defined above. Take three *AbMonoidExp* expressions, give the appropriate assignments and compute the results of evaluating the three expressions.

Exercise 4.26. A *ring* is a set A together with two constants, 0 and 1, one unary operation, *negate*, and two binary operations, $(+)$ and $(*)$, such that

a. 0 is the neutral element of $(+)$

$$\forall\, x \in A.\ x + 0 = 0 + x = x$$

b. $(+)$ is associative

$$\forall\, x, y, z \in A.\ x + (y + z) = (x + y) + z$$

c. *negate* inverts elements with respect to addition

$$\forall\, x \in A.\ x + negate\ x = negate\ x + x = 0$$

d. $(+)$ is commutative

$$\forall\, x, y \in A.\ x + y = y + x$$

e. 1 is the unit (neutral element) of $(*)$

$$\forall\, x \in A.\ x * 1 = 1 * x = x$$

f. $(*)$ is associative

$$\forall\, x, y, z \in A.\ x * (y * z) = (x * y) * z$$

g. $(*)$ distributes over $(+)$

$$\forall\, x, y, z \in A.\ x * (y + z) = (x * y) + (x * z)$$
$$\forall\, x, y, z \in A.\ (x + y) * z = (x * z) + (y * z)$$

Remarks:

- a. – b. say that $(A, 0, +)$ is a monoid and e. – f. that $(A, 1, *)$ is a monoid
- a. – c. say that $(A, 0, +, negate)$ is a group
- a. – d. say that $(A, 0, +, negate)$ is a commutative (abelian) group

i Define a type class *Ring* that corresponds to the ring structure.

ii Define a datatype for the language of ring expressions (including variables) and define a *Ring* instance for it.

iii Find two other instances of the *Ring* class.

iv Define a general evaluator for *Ring* expressions on the basis of a given assignment function (mapping variables to semantic values).

v Specialise the evaluator to the two *Ring* instances defined at point iii. Take three ring expressions, give the appropriate assignments and compute the results of evaluating, in each case, the three expressions.

Exercise 4.27. (Note that the *Num* hierarchy has been replaced by *Additive*, *AddGroup*, etc.)

Recall the type of expressions of one variable from Section 1.7.1.

```
data FunExp = Const Rational    | X
            | FunExp :+: FunExp  | Exp FunExp
            | FunExp :*: FunExp  | Sin FunExp
            | FunExp :/: FunExp  | Cos FunExp
            -- and so on
         deriving Show
```

and consider the function

$$f :: \mathbb{R} \to \mathbb{R}$$
$$f\ x = exp\ (sin\ x) + x$$

1. Find an expression e such that *eval e == f* and show this using equational reasoning.

2. Implement a function *deriv2* such that, for any $f :: Fractional\ a \Rightarrow a \to a$ constructed with the grammar of *FunExp* and any x in the domain of f, we have that *deriv2 f x* computes the second derivative of f at x. Use the function *derive :: FunExp → FunExp* from the lectures (*eval (derive e)* is the derivative of *eval e*). What instance declarations do you need?

   The type of *deriv2 f* should be *Fractional a ⇒ a → a*.

Exercise 4.28. Write a function *simplify :: FunExp → FunExp* to simplify the expression resulting from *derive*. For example, the following tests should work:

```
testSimplify =    -- all evaluate to True
  [simplify (Const 0 :*: Exp X)   == Const 0
  ,simplify (Const 0 :+: Exp X)   == Exp X
  ,simplify (Const 2 :*: Const 1) == Const 2
  ,simplify (derive (X :*: X))    == Const 2 :*: X
  ]
```

As a motivating example, note that without *simplify* we have *derive* $(X :*: X)$ == $(Const\ 1 :*: X) :+: (X :*: Const\ 1)$, and that the syntax tree of the second derivative is twice that size.

4.10.1   Project: Optimisation using Newton's method

This subsection describes a larger exercise (or small project) you can use to practice what you have learnt so far. It is heavily based on Chapter 3 and Chapter 4 (the *FunExp* type, *eval*, *derive*, *D*, tupling, homomorphisms, *FD*, *apply*, ...) so it pays off to work through those parts carefully.

Part 1   The evaluation of the second derivative is given by

$$eval'' = eval' \circ derive = eval \circ derive \circ derive$$

a) Let $P\ (h) = $ "h is a homomorphism from *FunExp* to *FunSem* $= \mathbb{R} \to \mathbb{R}$". Express P in logic and show $\neg\ P\ (eval'')$.

b) Given the types in the skeleton code below, define instances of the classes *Additive, AddGroup, Multiplicative, MulGroup,* and *Transcendental,* for *Tri a*. Test your results using algebraic identities like $sin^{\wedge}2 + cos^{\wedge}2 = const\ one$.

c) Define a homomorphism *evalDD* from *FunExp* to *FunTri a* (for any type a in the class *Transcendental*). You don't need to prove that it is a homomorphism in this part.

d) Show that *evalDD* is a homomorphism for the case of multiplication.

Part 2   Newton's method allows us to find zeros of a large class of functions in a given interval. The following description of Newton's method follows Bird and Wadler [1988], page 23:

$$newton :: (\mathbb{R} \to \mathbb{R}) \to \mathbb{R} \to \mathbb{R} \to \mathbb{R}$$
$$newton\ f\ \epsilon\ x = \textbf{if}\ abs\ fx < \epsilon \quad \textbf{then}\ x$$
$$\qquad\qquad\qquad \textbf{else if}\ fx' \neq 0\ \textbf{then}\ newton\ f\ \epsilon\ next$$
$$\qquad\qquad\qquad\qquad\qquad \textbf{else}\quad newton\ f\ \epsilon\ (x + \epsilon)$$
$$\quad\ \textbf{where}\ fx\ \ = f\ x$$
$$\qquad\qquad fx'\ \ = undefined \quad \text{-- should be}\ f'\ x\ (\text{derivative of}\ f\ \text{at}\ x)$$
$$\qquad\qquad next = x - (fx\ /\ fx')$$

a) Implement Newton's method, using $Tri\ \mathbb{R} \to Tri\ \mathbb{R}$ for the type of the first argument. In other words, use the code above to implement

$$newtonTri :: (Tri\ \mathbb{R} \to Tri\ \mathbb{R}) \to \mathbb{R} \to \mathbb{R} \to \mathbb{R}$$

in order to obtain the appropriate value for $f'\ x$.

b) Test your implementation on the following functions:

$test0\ x = x\char`^2$	-- one (double) zero, in zero
$test1\ x = x\char`^2 - one$	-- two zeros, in +-one
$test2\ x = sin\ x$	-- many, many zeros (in $n * \pi$ for all $n :: \mathbb{Z}$)
$test3\ n\ x\ y = y\char`^n - constTri\ x$	-- $test3\ n\ x$, has zero in "nth roots of x"
-- where $constTri$ is the embedding of $Const$	

Note that these functions can work at different types: $\mathbb{R} \to \mathbb{R}$ or $Dup\ \mathbb{R} \to Dup\ \mathbb{R}$ or $Tri\ \mathbb{R} \to Tri\ \mathbb{R}$, etc.

For each of these functions, apply Newton's method to a number of starting points from a sensible interval. For example:

$$map\ (newton\ test1\ 0.001)\ [-2.0, -1.5 .. 2.0]$$

but be aware that the method might not always converge!

For debugging is advisable to implement *newton* in terms of *newtonList*, a minor variation which returns a list of the approximations encountered on the way to the final answer:

$$newton\ f\ \epsilon\ x = last\ (newtonList\ f\ \epsilon\ x)$$
$$newtonList\ f\ \epsilon\ x = x : \textbf{if} \dots \textbf{then}\ [\ ]\ \textbf{else} \dots$$

Part 3 We can find the optima of a twice-differentiable function on an interval by finding the zeros of its derivative on that interval, and checking the second derivative. If $f'\ x_0$ is zero, then

- if $f''\ x_0 < 0$, then x_0 is a maximum

- if $f''\ x_0 > 0$, then x_0 is a minimum

- if $f''\ x_0 = 0$, then, if $f''\ (x_0 - \epsilon) * f''\ (x_0 + \epsilon) < 0$ (i.e., f'' changes its sign in the neighbourhood of x_0), x_0 is an inflection point (not an optimum)

- otherwise, we don't know

Use Newton's method to find the optima of the test functions from point 2. That is, implement a function

$$optim :: (Tri\ \mathbb{R} \to Tri\ \mathbb{R}) \to \mathbb{R} \to \mathbb{R} \to Result\ \mathbb{R}$$

so that $optim\ f\ \epsilon\ x$ uses Newton's method to find a zero of f' starting from x. If y is the result (i.e. $f'\ y$ is within ϵ of 0), then check the second derivative, returning $Maximum\ y$ if $f''\ y < 0$, $Minimum\ y$ if $f''\ y > 0$, and $Dunno\ y$ if $f'' = 0$.

As before, use several starting points to test if you get the expected behaviour.

Skeleton code   Here is some useful skeleton Haskell code to start from, and the *Algebra* and *FunExp* modules are also available on github.

```
{-# LANGUAGE FlexibleContexts, FlexibleInstances #-}
{-# LANGUAGE TypeSynonymInstances #-}
module A2_Skeleton where
import Prelude hiding ((+), (−), (∗), (/), negate, recip, (^),
                        π, sin, cos, exp, fromInteger, fromRational)
import DSLsofMath.Algebra
import DSLsofMath.FunExp

type Tri a    = (a, a, a)
type TriFun a = Tri (a → a)   -- = (a → a, a → a, a → a)
type FunTri a = a → Tri a     -- = a → (a, a, a)

instance Additive a ⇒ Additive (Tri a) where
  (+) = addTri; zero    = zeroTri
instance (Additive a, Multiplicative a) ⇒ Multiplicative (Tri a) where
  (∗)  = mulTri; one    = oneTri

instance AddGroup a ⇒ AddGroup (Tri a) where
  negate = negateTri
instance (AddGroup a, MulGroup a) ⇒ MulGroup (Tri a) where
  recip  = recipTri

(addTri, zeroTri, mulTri, oneTri, negateTri, recipTri) = undefined

instance Transcendental a ⇒ Transcendental (Tri a) where
  π = piTri; sin = sinTri; cos = cosTri; exp = expTri

(piTri, sinTri, cosTri, expTri) = undefined
```

Chapter 5

Polynomials and Power Series

5.1 Polynomials

Again we take as starting point a definition from Adams and Essex [2010], this time from page 39:

> A **polynomial** is a function P whose value at x is
>
> $$P(x) = a_n x^n + a_{n-1} x^{n-1} + \cdots + a_1 x + a_0$$
>
> where $a_n, a_{n-1}, \ldots, a_1$, and a_0, called the **coefficients** of the polymonial [sic], are constants and, if $n > 0$, then $a_n \neq 0$. The number n, the degree of the highest power of x in the polynomial, is called the **degree** of the polynomial. (The degree of the zero polynomial is not defined.)

This definition raises a number of questions, for example "what is the zero polynomial?" (and why isn't its degree defined).

The types of the elements involved in the definition appear to be

$$P : \mathbb{R} \to \mathbb{R}; \ x : \mathbb{R}; \ n : \mathbb{N}; \ a_0, \ldots, a_n : \mathbb{R} \text{ with } a_n \neq 0 \text{ if } n > 0$$

141

The phrasing should be "whose value at *any* x is". The remark that the a_i are constants is probably meant to indicate that they do not depend on x, otherwise every function would be a polynomial. The zero polynomial is, according to this definition, the *const* 0 function. Thus, what is meant is

> A **polynomial** is a function $P : \mathbb{R} \to \mathbb{R}$ which either is the constant zero function, or there exist $a_0, \ldots, a_n : \mathbb{R}$ with $a_n \neq 0$ (called **coefficients**) such that, for every $x : \mathbb{R}$
>
> $$P(x) = a_n x^n + a_{n-1} x^{n-1} + \cdots + a_1 x + a_0$$
>
> For the constant zero polynomial the degree is not defined. Otherwise, the degree is n.

Syntax and semantics of polynomials   Given the coefficients a_i we can evaluate P at any given x. Ignoring the condition on coefficients for now, we can assume that the coefficients are given as a list $as = [a_0, a_1, \ldots, a_n]$ (we prefer counting up). Then the evaluation function is:

```
evalL :: [ℝ] → ℝ → ℝ
evalL  []        x  = 0
evalL  (a : as)  x  = a + x * evalL as x
```

Note that we can read the type as $evalL :: [\mathbb{R}] \to (\mathbb{R} \to \mathbb{R})$ and thus identify $[\mathbb{R}]$ as the type for the (abstract) syntax (for polynomials) and $(\mathbb{R} \to \mathbb{R})$ as the type of the semantics (for polynomial functions). A good exercise (5.2) is to show that this evaluation function gives the same result as the formula above.

Using the *Ring* instance for functions we can rewrite *eval* into a one-argument function (returning a polynomial function):

```
evalL :: [ℝ] → (ℝ → ℝ)
evalL []       = const 0
evalL (a : as) = const a + id * evalL as
```

As an example, the polynomial which is usually written just x is represented by the list $[0, 1]$ and the polynomial function $\lambda x \to x^2 - 1$ is represented by the list $[-1, 0, 1]$.

It is worth noting that the definition of what we call a "polynomial function" is semantic, not syntactic. A syntactic definition would talk about the form of the expression (a sum of coefficients times natural powers of x). In contrast, this semantic definition only requires that the function P *behaves like* such a sum. Insisting on this difference may seem pedantic, but here is an interesting example of a family of functions which syntactically does not look like a sum of powers:

$$T_n(x) = \cos(n * \arccos(x)).$$

And yet, it can be shown that T_n is a polynomial function of degree n (on the interval $[-1, 1]$). Exercise 5.4 guides you to a proof. At this point you could just compute T_0, T_1, and T_2 by hand to get a feeling for how it works.

Not every list of coefficients is valid according to the definition. In particular, the empty list is not a valid list of coefficients, so we have a conceptual, if not empirical, type error in our evaluator.

The valid lists are those *finite* lists in the set

$$\{[0]\} \cup \{(a : as) \mid last\ (a : as) \neq 0\}$$

The fact that the element should be non-zero is easy to express as a Haskell expression (*last* $(a : as) \neq 0$), but not so easy to express in the *types*.

We could try jumping through the relevant hoops. However, at this stage, we can realise that the non-zero condition is there only to define the degree of the polynomial. The same can be said about the separation between zero and non-zero polynomials, which is there to explicitly leave the degree undefined. So we can further improve the definition as follows:

A **polynomial** is a function $P : \mathbb{R} \to \mathbb{R}$ such that there exist $a_0, \ldots,$ $a_n : \mathbb{R}$ and for any $x : \mathbb{R}$

$$P(x) = a_n x^n + a_{n-1} x^{n-1} + \cdots + a_1 x + a_0$$

The degree of the polynomial is the largest i such that $a_i \neq 0$.

This definition is much simpler to manipulate and clearly separates the definition of degree from the definition of polynomial. Perhaps surprisingly, there is no longer any need to single out the zero polynomial to define the degree. Indeed, when the polynomial is zero, $a_i = 0$ for every i, and we have an empty set of indices where $a_i \neq 0$. The largest element of this set is undefined (by definition of largest, see also Exercise 5.11), and we have the intended definition.

Representing polynomials So, we can simply use any list of coefficients to *represent* a polynomial:

newtype *Poly a* $=$ *Poly* $[a]$ **deriving** (*Show, Eq*)

Since we only use the arithmetic operations, we can generalise our evaluator to an arbitrary *Ring* type.

$$evalPoly :: Ring\ a \Rightarrow Poly\ a \to (a \to a)$$
$$evalPoly\ (Poly\ [])\qquad \_ = 0$$
$$evalPoly\ (Poly\ (a:as))\ x = a + x * evalPoly\ (Poly\ as)\ x$$

Since we have *Ring a*, there is a *Ring* structure on $a \to a$, and *evalPoly* looks like a homomorphism. Question: is there a *Ring* structure on *Poly a*, such that *evalPoly* is a homomorphism?

For example, the homomorphism condition gives for $(+)$

$$evalPoly\ as + evalPoly\ bs = evalPoly\ (as + bs)$$

Note that this equation uses $(+)$ at two different type: on the left hand side (lhs) two functions of type $a \to a$ are added (pointwise) and on the right hand side (rhs) two *Poly a* (lists of coefficients) are added. We are using the homomorphism condition to find requirements on the definition of $(+)$ on *Poly a*.

Both sides (lhs and rhs) are functions, thus they are equal if and only if they are equal for every argument. For an arbitrary x

$$(evalPoly\ as + evalPoly\ bs)\ x = evalPoly\ (as + bs)\ x$$
$$\Leftrightarrow \{- (+)\ \text{on functions is defined point-wise} -\}$$
$$evalPoly\ as\ x + evalPoly\ bs\ x = evalPoly\ (as + bs)\ x$$

To proceed further, we need to consider the various cases in the definition of *evalPoly* and use list induction. We give the computation for the step case, dropping the *Poly* constructor by using $eval\ cs = evalPoly\ (Poly\ cs)$ for brevity.

$$eval\ (a:as)\ x + eval\ (b:bs)\ x = eval\ ((a:as) + (b:bs))\ x$$

We use the homomorphism condition for *as* and *bs*. For the left-hand side, we have:

$$
\begin{aligned}
eval\ (a:as)\ x + eval\ (b:bs)\ x &\quad = \{-\ \text{def. } eval\ -\}\\
(a + x * eval\ as\ x) + (b + x * eval\ bs\ x) &\quad = \{-\ \text{arithmetic (ring laws)}\ -\}\\
(a + b) + x * (eval\ as\ x + eval\ bs\ x) &\quad = \{-\ \text{homomorphism condition}\ -\}\\
(a + b) + x * (eval\ (as + bs)\ x) &\quad = \{-\ \text{def. } eval\ -\}\\
eval\ ((a + b) : (as + bs))\ x
\end{aligned}
$$

The homomorphism condition will hold for every x if we define

$$(a:as) + (b:bs) = (a + b) : (as + bs)$$

This definition looks natural (we could probably have guessed it early on) but it is still interesting to see that we can derive it as the form that it has to take for the proof to go through.

Numeric instances for polynomials We leave the derivation of the other cases and operations as an exercise. Here, we just give the corresponding definitions.

> **instance** *Additive a* $\Rightarrow$ *Additive* (*Poly a*) **where**
>   (+) = *addPoly*; *zero* = *Poly* [ ]
> *addPoly* :: *Additive a* $\Rightarrow$ *Poly a* $\rightarrow$ *Poly a* $\rightarrow$ *Poly a*
> *addPoly* (*Poly xs*) (*Poly ys*) = *Poly* (*addList xs ys*)
> *addList* :: *Additive a* $\Rightarrow$ [*a*] $\rightarrow$ [*a*] $\rightarrow$ [*a*]
> *addList* = *zipWithLonger* (+)
> *zipWithLonger* :: (*a* $\rightarrow$ *a* $\rightarrow$ *a*) $\rightarrow$ ([*a*] $\rightarrow$ [*a*] $\rightarrow$ [*a*])
> *zipWithLonger* _ [ ]   *bs*   = *bs*   -- 0 + *bs* == *bs*
> *zipWithLonger* _ *as*   [ ]   = *as*   -- *as* + 0 == *as*
> *zipWithLonger op* (*a* : *as*) (*b* : *bs*) = *op a b* : *zipWithLonger op as bs*
> **instance** *AddGroup a* $\Rightarrow$ *AddGroup* (*Poly a*) **where**
>   *negate* = *negPoly*
> *negPoly* :: *AddGroup a* $\Rightarrow$ *Poly a* $\rightarrow$ *Poly a*
> *negPoly* = *polyMap negate*
> *polyMap* :: (*a* $\rightarrow$ *b*) $\rightarrow$ (*Poly a* $\rightarrow$ *Poly b*)
> *polyMap f* (*Poly as*) = *Poly* (*map f as*)
> **instance** *Ring a* $\Rightarrow$ *Multiplicative* (*Poly a*) **where**
>   (*) = *mulPoly*; *one* = *Poly* [*one*]
> *mulPoly* :: *Ring a* $\Rightarrow$ *Poly a* $\rightarrow$ *Poly a* $\rightarrow$ *Poly a*
> *mulPoly* (*Poly xs*) (*Poly ys*) = *Poly* (*mulList xs ys*)
> *mulList* :: *Ring a* $\Rightarrow$ [*a*] $\rightarrow$ [*a*] $\rightarrow$ [*a*]
> *mulList* [ ]   _   = [ ]   -- 0 * *bs* == 0
> *mulList* _   [ ]   = [ ]   -- *as* * 0 == 0
> *mulList* (*a* : *as*) (*b* : *bs*) = (*a* * *b*) : *addList* (*scaleList a bs*)
>   (*mulList as* (*b* : *bs*))
> *scaleList* :: *Multiplicative a* $\Rightarrow$ *a* $\rightarrow$ [*a*] $\rightarrow$ [*a*]
> *scaleList a* = *map* (*a**)

As we *can* define a *Ring* structure on *Poly a*, and we have arrived at the canonical definition of polynomials, as found in any algebra book (see, for example, Rotman [2006] for a very readable text):

> Given a commutative ring *A*, the commutative ring given by the set *Poly A* together with the operations defined above is the ring of **polynomials** with coefficients in *A*.

Note that from here on we will use the term "polynomial" for the abstract syntax (the list of coefficients, *as*) and "polynomial function" for its semantics (the function *evalPoly as* : $A \rightarrow A$).

An alternative representation   The canonical representation of polynomials in algebra does not use finite lists, but the equivalent

$Poly' A = \{ a : \mathbb{N} \rightarrow A \mid \{\text{-} a \text{ has a finite number of non-zero values -}\} \}$

Exercise 5.7: What are the ring operations on *Poly' A*? For example, here is the specification of addition:

$a + b = c \Leftrightarrow \forall n : \mathbb{N}. \ a \ n + b \ n = c \ n$

Hint: they are not all the same as the operations on arbitrary functions $X \rightarrow A$ defined in Section 3.5.3.

Remark: Using functions from $\mathbb{N}$ in the definition has certain technical advantages over using finite lists. For example, consider adding $[a_0, a_1, \ldots, a_n]$ and $[b_0, b_1, \ldots, b_m]$, where $n > m$. Then, we obtain a polynomial of degree n: $[c_0, c_1, \ldots, c_n]$. The formula for the c_i must now be given via a case distinction:

$c_i = $ **if** $i > m$ **then** a_i **else** $a_i + b_i$

since b_i does not exist for values greater than m.

Compare this with the formula for functions from $\mathbb{N}$, where no case distinction is necessary. The advantage is even clearer in the case of multiplication.

Syntax $\neq$ semantics   If one considers arbitrary rings, polynomials are not isomorphic (in one-to-one correspondence) to polynomial functions. For any finite ring A, there is a finite number of functions $A \rightarrow A$, but there is a countable infinity of polynomials. That means that the same polynomial function on A will be the evaluation of many different polynomials.

For example, consider the ring $\mathbb{Z}_2$ ($\{0, 1\}$ with addition and multiplication modulo 2). In this ring, we have that $p \ x = x + x\texttt{\^{}}2$ is actually a constant function. The only two input values to p are 0 and 1 and we can easily check that $p \ 0 = 0$ and also $p \ 1 = (1 + 1\texttt{\^{}}2)\%2 = 2\%2 = 0$. Thus

$eval \ [0, 1, 1] = p = const \ 0 = eval \ [\,]$ -- in $\mathbb{Z}_2 \rightarrow \mathbb{Z}_2$

but

$[0, 1, 1] \neq [\,]$ -- in *Poly* $\mathbb{Z}_2$

Therefore, it is not generally a good idea to conflate polynomials (syntax) and polynomial functions (semantics).

Algebra of syntactic polynomials   Following the DSL terminology, we can say that the polynomial functions are the semantics of the language of polynomials. We started with polynomial functions, we wrote the evaluation function and realised that we have the makings of a homomorphism. That suggested that we could create an adequate language for polynomial functions. Indeed, this turns out to be the case; in so doing, we have recreated an important mathematical achievement: the algebraic definition of polynomials.

Let

$$x :: Ring\ a \Rightarrow Poly\ a$$
$$x = Poly\ [0, 1]$$

Then for any polynomial $as = Poly\ [a_0, a_1, \ldots, a_n]$ we have

$$as = a_0 \cdot x\char`^0 + a_1 \cdot x\char`^1 + a_2 \cdot x\char`^2 + \ldots + a_n \cdot x\char`^n$$

where $(+)$ is addition of coefficient lists and $(\cdot)$ is an infix version of *scaleList*. Exercise 5.3: Prove the above equality.

This equality justifies the standard notation

$$as \mathrel{==} \sum_{i=0}^{n} a_i \cdot x\char`^i$$

where both sides of the equality are syntax (expressions of type *Poly a*).

5.2   Division and the degree of the zero polynomial

Recall the fundamental property of division that we learned in high school:

> For all naturals a, b, with $b \neq 0$, there exist *unique* integers q and r, such that
>
> $$a = b * q + r, \qquad \text{with} \quad 0 \leqslant r < b$$
>
> When $r = 0$, a is divisible by b.

Questions of divisibility are essential in number theory and its applications (including cryptography). A similar theorem holds for polynomials (see, for example, [Adams and Essex, 2010, page 40]):

> For all polynomials as, bs, with $bs \neq 0$, there exist *unique* polynomials qs and rs, such that
>
> $$as = bs * qs + rs, \qquad \text{with} \quad degree\ rs < degree\ bs$$

The condition $r < b$ is replaced by *degree rs < degree bs*. However, we now have a problem. Every polynomial is divisible by any non-zero constant polynomial, resulting in a zero polynomial remainder. But the degree of a constant polynomial is zero. If the degree of the zero polynomial were a natural number, it would have to be smaller than zero. For this reason, it is either considered undefined (as in Adams and Essex [2010]), or it is defined as $-\infty$.[1] The next section examines this question from the point of view of homomorphisms.

5.3　Polynomial degree as a homomorphism

It is often the case that a certain function is *almost* a homomorphism and the domain or range structure is *almost* a monoid. In Section 4.6, we saw tupling as one way to fix such a problem and here we will introduce another way.

The *degree* of a polynomial is a good candidate for being a homomorphism: if we multiply two polynomials we can normally add their degrees. If we try to check that *degree* :: *Poly a* $\to$ $\mathbb{N}$ is the function underlying a monoid morphism we need to decide on the monoid structure to use for the source and for the target, and we need to check the homomorphism laws. We can use the multiplicative monoid (*unit* = *one* and *op* = *mulPoly*) for the source and we can try to use the additive monoid (*unit* = *zero* and *op* = (+)) for the target monoid. Then we need to check that

> *degree one* = *zero*
> $\forall\, x, y.\ degree\ (x\ 'op'\ y) = degree\ x + degree\ y$

The first law is no problem and for most polynomials the second law is also straightforward to prove (try it as an exercise). But we run into trouble with one special case: the zero polynomial.

Looking back at the definition from [Adams and Essex, 2010, page 55] it says that the degree of the zero polynomial is not defined. Let's see why that is the case and how we might "fix" it. Assume that there exists a natural number z such that *degree* $0 = z$ and let $p = x\hat{\ }n$ so that *degree* $p = n$. Then we get

> z　　　　　　　　　　　 = {- assumption -}
> *degree* 0　　　　　　　 = {- simple calculation -}
> *degree* $(0 * p)$　　　　 = {- homomorphism condition -}
> *degree* $0 + degree\ p$　 = {- assumption -}
> $z + n$

---

[1]Likewise we could define the largest element of the empty set to be $-\infty$.

Thus we need to find a degree z, for the zero polynomial, such that $z = z + n$ for all natural numbers n! At this stage we could either give up, or think out of the box. Intuitively we could try to use $z = -Infinity$, which would seem to satisfy the law but is not a natural number (not even an integer). More formally what we need to do is to extend the monoid $(\mathbb{N}, 0, +)$ by one more element. In Haskell we can do that using the *Maybe* type constructor:

```
class Monoid a where
  unit :: a
  op   :: a → a → a
instance Monoid a ⇒ Monoid (Maybe a) where
  unit = Just unit
  op   = opMaybe
opMaybe :: Monoid a ⇒ Maybe a → Maybe a → Maybe a
opMaybe Nothing   _m        = Nothing   -- (−Inf) + m = −Inf
opMaybe _m        Nothing   = Nothing   -- m + (−Inf) = −Inf
opMaybe (Just m₁) (Just m₂) = Just (op m₁ m₂)
```

Thus, to sum up, *degree* is a monoid homomorphism from $(Poly\ a, 1, *)$ to $(Maybe\ \mathbb{N}, Just\ 0, opMaybe)$.

Exercise 5.9: Check all the Monoid and homomorphism properties.

5.4 Power Series

Consider the following (false) proposition:

Proposition 5.1. Let $m, n \in \mathbb{N}$ and let *cs* and *as* be any polynomials of degree $m + n$ and n, respectively, and with $a_0 \neq 0$. Then there exists a polynomial *bs* of degree m such that $cs = as * bs$ (thus *cs* is divisible by *as*).

Even if the proposition is false, we can make the following proof attempt:

Proof. We need to find $bs = [b_0, \ldots, b_m]$ such that $cs = as * bs$. From the multiplication of polynomials, we know that

$$c_k = \sum_{i=0}^{k} a_i * b_{k-i}$$

Therefore:

$$c_0 = a_0 * b_0$$

Since c_0 and a_0 are known, computing $b_0 = c_0 \,/\, a_0$ is trivial. Next

$$c_1 = a_0 * b_1 + a_1 * b_0$$

Again, we are given c_1, a_0 and a_1, and we have just computed b_0, therefore we can obtain $b_1 = (c_1 - a_1 * b_0) \,/\, a_0$. Similarly

$$c_2 = a_0 * b_2 + a_1 * b_1 + a_2 * b_0$$

from which we obtain, as before, the value of b_2 by subtraction and division.

It is clear that this process can be continued, yielding at every step a value for a coefficient of bs, and thus we have obtained bs satisfying $cs = as * bs$.   □

The problem with this proof attempt is in the statement "it is clear that this process can be continued". In fact, it is rather clear that it cannot be continued (for polynomials)! Indeed, bs only has $m + 1$ coefficients, therefore for all remaining n equations of the form $c_k = \sum_{i=0}^{k} a_i * b_{k-i}$, the values of b_k (for $k > m$) have to be zero. But in general this will not satisfy the equations.

However, we can now see that, if we were able to continue forever, we would be able to divide cs by as exactly. The only obstacle is the "finite" nature of our lists of coefficients.

Power series are obtained from polynomials by removing in *Poly'* the restriction that there should be a *finite* number of non-zero coefficients; or, in, the case of *Poly*, by going from finite lists to infinite streams.

    type *PowerSeries a = Poly a*   -- finite and infinite lists

The operations are still defined as before. If we consider only infinite lists, then only the equations which deal with non-empty lists will apply.

Power series are usually denoted

$$\sum_{n=0}^{\infty} a_n * x^n$$

the interpretation of x being the same as before. The simplest operation, addition, can be illustrated as follows:

$$\sum_{i=0}^{\infty} a_i * x^i \quad\cong\quad [a_0, \qquad a_1, \qquad \ldots]$$

$$\sum_{i=0}^{\infty} b_i * x^i \quad\cong\quad [b_0, \qquad b_1, \qquad \ldots]$$

$$\sum_{i=0}^{\infty} (a_i + b_i) * x^i \quad\cong\quad [a_0 + b_0, \quad a_1 + b_1, \quad \ldots]$$

The evaluation of a power series represented by $a : \mathbb{N} \to A$ is defined, in case the necessary operations make sense on A, as a function

$$eval\ a : A \to A$$
$$eval\ a\ x = lim\ s\ \textbf{where}\ s\ n = \sum_{i=0}^{n} a_i * x^i$$

We will focus on the case in which $A = \mathbb{R}$ or $A = \mathbb{C}$. Note that $eval\ a$ is, in general, a partial function (the limit might not exist). To make $eval\ a$ a total function, the domain A would have to be restricted to just those values of x for which the limit exists (the infinite sum converges). Keeping track of different domains for different power series is cumbersome and the standard treatment is to work with "formal power series" with no requirement of convergence.

Here the qualifier "formal" refers to the independence of the definition of power series from the ideas of convergence and evaluation. In particular, two power series represented by a and b, respectively, are equal only if $a = b$ (as infinite series of numbers). If $a \neq b$, then the power series are different, even if $eval\ a = eval\ b$.

Since we cannot in general compute limits, we can use an approximative $eval$, by evaluating the polynomial initial segment of the power series.

$$evalPS :: Ring\ a \Rightarrow Int \to PowerSeries\ a \to (a \to a)$$
$$evalPS\ n\ as = evalPoly\ (takePoly\ n\ as)$$
$$takePoly :: Int \to PowerSeries\ a \to Poly\ a$$
$$takePoly\ n\ (Poly\ xs) = Poly\ (take\ n\ xs)$$

Note that $evalPS\ n$ is *not* a homomorphism: for example:

$$
\begin{aligned}
evalPS\ 2\ (x * x)\ 1 &=\\
evalPoly\ (takePoly\ 2\ [0,0,1])\ 1 &=\\
evalPoly\ [0,0]\ 1 &=\\
0
\end{aligned}
$$

but

$$
\begin{aligned}
(evalPS\ 2\ x\ 1) &=\\
evalPoly\ (takePoly\ 2\ [0,1])\ 1 &=\\
evalPoly\ [0,1]\ 1 &=\\
1
\end{aligned}
$$

and thus $evalPS\ 2\ (x * x)\ 1 = 0 \neq 1 = 1 * 1 = (evalPS\ 2\ x\ 1) * (evalPS\ 2\ x\ 1)$.

5.5   Operations on power series

Power series have a richer structure than polynomials. For example, as suggested above, we also have division (this is reminiscent of the move from $\mathbb{Z}$ to $\mathbb{Q}$ to allow division to be generalised). To illustrate, let us start with a special case: trying to compute $p = \frac{1}{1-x}$ as a power series. The specification of $a \mathbin{/} b = c$ is $a = c * b$, thus in our case we need to find a p such that $1 = (1-x) * p$. For polynomials there is no solution to this equation. One way to see that is by using the homomorphism *degree*: the degree of the left hand side is 0 and the degree of the RHS is $1 + degree\ p \neq 0$. But there is still hope if we move to formal power series.

Remember that p is then represented by a stream of coefficients, and let that stream be $[p_0, p_1, \ldots]$. We make a table of the coefficients of the RHS $= (1 - x) * p = p - x * p$ and of the LHS $= 1$ (seen as a power series).

$$
\begin{array}{lll}
p & \mathrel{==} [p_0, p_1, & p_2, & \cdots \\
x * p & \mathrel{==} [0,\ p_0, & p_1, & \cdots \\
p - x * p & \mathrel{==} [p_0, p_1 - p_0, p_2 - p_1, \cdots \\
1 & \mathrel{==} [1,\ 0, & 0, & \cdots
\end{array}
$$

Thus, to make the last two lines equal, we are looking for coefficients satisfying $p_0 = 1$, $p_1 - p_0 = 0$, $p_2 - p_1 = 0$, ... The solution is unique: $1 = p_0 = p_1 = p_2 = \ldots$ but it only exists for streams (infinite lists) of coefficients. In the common maths notation we have just computed

$$
\frac{1}{1-x} = \sum_{i=0}^{\infty} x^i
$$

Note that this equation holds when we interpret both sides as formal power series, but not necessarily if we try to evaluate the expressions for a particular x. Indeed, the RHS will converge if $|x| < 1$ but not for $x = 2$, for example.

For a more general case of power series division, consider $p \mathbin{/} q$ with $p = a : as$, $q = b : bs$, and assume that $a * b \neq 0$. Then we want to find, for any given $(a : as)$ and $(b : bs)$, the series $(c : cs)$ satisfying

$$
\begin{array}{ll}
(a : as) \mathbin{/} (b : bs) = (c : cs) & \Leftrightarrow \{\text{- spec. of division -}\} \\
(a : as) = (c : cs) * (b : bs) & \Leftrightarrow \{\text{- def. of } * \text{ for } Cons \text{ -}\} \\
(a : as) = (c * b) : ([c] * bs + cs * (b : bs)) \Leftrightarrow \{\text{- equality on components -}\} \\
a = c * b \qquad\qquad \{\text{- and -}\} \\
as = [c] * bs + cs * (b : bs) & \Leftrightarrow \{\text{- solve for } c \text{ and } cs \text{ -}\} \\
c = a \mathbin{/} b \qquad\qquad \{\text{- and -}\} \\
cs = (as - [c] * bs) \mathbin{/} (b : bs)
\end{array}
$$

This leads to the implementation:

> **instance** *(Eq a, Field a)* $\Rightarrow$ *MulGroup (PowerSeries a)* **where**
> $\quad$ $(/) = divPS$
>
> $divPS :: (Eq\ a, Field\ a) \Rightarrow PowerSeries\ a \rightarrow PowerSeries\ a \rightarrow PowerSeries\ a$
> $divPS\ (Poly\ as)\ (Poly\ bs) = Poly\ (divL\ as\ bs)$
>
> $divL :: (Eq\ a, Field\ a) \Rightarrow [a] \rightarrow [a] \rightarrow [a]$
> $divL\ [\,]\quad\_bs\quad = [\,]$ $\qquad\qquad$ -- case 0 / q
> $divL\ (0:as)\ (0:bs) = divL\ as\ bs$ $\qquad$ -- case xp / xq
> $divL\ (0:as)\ bs\quad = 0:divL\ as\ bs$ $\quad$ -- case xp / q
> $divL\ as\quad\quad [b]\quad = scaleList\ (1\ /\ b)\ as$ $\ $ -- case p / c
> $divL\ (a:as)\ (b:bs) = c:divL\ (addList\ as\ (scaleList\ (-c)\ bs))\ (b:bs)$
> $\qquad\qquad\qquad\qquad$ **where** $c = a\ /\ b$
> $divL\ \_\quad\quad [\,]\quad\ = error\ \texttt{"divL: division by zero"}$

This definition allows us to also use division on polynomials, but the result will, in general, be a power series, not a polynomial. The different cases can be calculated from the specification. Some examples:

> $ps_0, ps_1, ps_2 :: (Eq\ a, Field\ a) \Rightarrow PowerSeries\ a$
> $ps_0 = 1\ /\ (1-x)$ $\qquad\qquad$ -- ps_0 == $Poly\ [1,1,1,1,\dots]$
> $ps_1 = 1\ /\ (1-x)\hat{\ }2$ $\qquad\quad$ -- ps_1 == $Poly\ [1,2,3,4,\dots]$
> $ps_2 = (x\hat{\ }2 - 2*x + 1)\ /\ (x-1)$ $\ $ -- ps_2 == $Poly\ [-1,1,0]$

Every *ps* is the result of a division of polynomials: the first two return power series, the third is a polynomial (even though it ends up having a trailing zero). We can get a feeling for the definition by computing ps_0 "by hand". We let $p = [1]$ and $q = [1, -1]$ and seek $r = p\ /\ q$.

> $divL\ p\ q$ $\qquad\qquad\qquad\qquad\qquad\qquad$ = {- def. of p and q -}
> $divL\ (1:[\,])\ (1:[-1])$ $\qquad\qquad\qquad\quad$ = {- main case of *divL* -}
> $(1\ /\ 1):divL\ ([\,] - [1] * [-1])\ (1:[-1])$ = {- simpl., def. of $(*)$, $(-)$ -}
> $1:divL\ [1]\ (1:[-1])$ $\qquad\qquad\qquad\quad$ = {- def. of p and q -}
> $1:divL\ p\ q$

Thus, the answer *r* starts with 1 and continues with *r*! In other words, we have that $1\ /\ [1, -1] = [1, 1 ..]$ as an infinite list of coefficients and $\frac{1}{1-x} = \sum_{i=0}^{\infty} x^i$ in the more traditional mathematical notation.

5.6 Formal derivative

Considering the analogy between power series and polynomial functions (via polynomials), we can arrive at a formal derivative for power series through the following computation:

$$\left(\sum_{n=0}^{\infty} a_n * x^n \right)' = \sum_{n=0}^{\infty} (a_n * x^n)' = \sum_{n=0}^{\infty} a_n * (x^n)' = \sum_{n=0}^{\infty} a_n * (n * x^{n-1})$$

$$= \sum_{n=0}^{\infty} (n * a_n) * x^{n-1} = \sum_{n=1}^{\infty} (n * a_n) * x^{n-1} \qquad (5.1)$$

$$= \sum_{m=0}^{\infty} ((m+1) * a_{m+1}) * x^m$$

Thus the mth coefficient of the derivative is $(m+1) * a_{m+1}$.

We can implement this formula, for example, as

> *deriv* :: *Ring a* $\Rightarrow$ *Poly a* $\rightarrow$ *Poly a*
> *deriv* (*Poly as*) = *Poly* (*derivL as*)
> *derivL* :: *Ring a* $\Rightarrow$ [*a*] $\rightarrow$ [*a*]
> *derivL* [ ]   = [ ]
> *derivL* (_ : *as*) = *zipWith* (*) *oneUp as*
> *oneUp* :: *Ring a* $\Rightarrow$ [*a*]
> *oneUp* = *one* : *map* (*one*+) *oneUp*

Side note: we cannot in general implement a decidable (Boolean) equality test for *PowerSeries*. For example, we know that *deriv* ps_0 equals ps_1 but we cannot compute *True* in finite time by comparing the coefficients of the two power series.

> *checkDeriv* :: *Int* $\rightarrow$ *Bool*
> *checkDeriv n* = *takePoly n* (*deriv* ps_0) == *takePoly n* (ps_1 :: *Poly Rational*)

Recommended reading: the Functional pearl "Power series, power serious" McIlroy [1999].

5.7 Exercises

The first few exercises are about filling in the gaps in the chapter above.

Exercise 5.1. Polynomial multiplication. To get a feeling for the definition it can be useful to take it step by step, starting with some easy cases.

$$mulP \; [] \; p = \quad \text{-- TODO}$$
$$mulP \; p \; [] = \quad \text{-- TODO}$$

$$mulP \; [a] \; p = \quad \text{-- TODO}$$
$$mulP \; p \; [b] = \quad \text{-- TODO}$$

$$mulP \; (0:as) \; p = \quad \text{-- TODO}$$
$$mulP \; p \; (0:bs) = \quad \text{-- TODO}$$

Finally we reach the main case

$$mulP \; (a:as) \; q@(b:bs) = \quad \text{-- TODO}$$

Exercise 5.2. Show (by induction) that the evaluation function *evalL* (from page 142) gives the same result as the formula

$$P(x) = a_n x^n + a_{n-1} x^{n-1} + \cdots + a_1 x + a_0$$

from the quote on that same page.

Exercise 5.3. Prove that, with the definitions $x = [0, 1]$ and $as = [a_0, a_1, \ldots, a_n]$, we really have

$$as = a_0 \cdot x\char`^0 + a_1 \cdot x\char`^1 + a_2 \cdot x\char`^2 + \ldots + a_n \cdot x\char`^n$$

where $(+)$ is addition of coefficient lists and $(\cdot)$ is an infix version of *scaleList*.

Exercise 5.4. Chebyshev polynomials. Let $T_n(x) = \cos(n * \arccos(x))$ for x in the interval $[-1, 1]$. Compute T_0, T_1, and T_2 by hand to get a feeling for how it works. Note that they all turn out to be polynomial functions. In fact, T_n is a polynomial function of degree n for all n. To prove this, here are a few hints:

- $\cos(\alpha) + \cos(\beta) = 2 * \cos(\frac{\alpha+\beta}{2}) * \cos(\frac{\alpha-\beta}{2})$
- let $\alpha = (n+1) * \arccos(x)$ and $\beta = (n-1) * \arccos(x)$
- Simplify $T_{n+1}(x) + T_{n-1}(x)$ to relate it to $T_n(x)$.

- Note that the relation can be seen as an inductive definition of $T_{n+1}(x)$.

- Use induction on n.

Exercise 5.5. Another view of T_n from Exercise 5.4 is as a homomorphism. Let $H_1(h, F, f) = \forall x. \; h \; (F \; x) = f \; (h \; x)$ be the predicate that states "$h : A \to B$ is a homomorphism from $F : A \to A$ to $f : B \to B$". Show that $H_1(cos, (n*), T_n)$ holds, where $cos : \mathbb{R}_{\geq 0} \to [-1, 1]$, $(n*) : \mathbb{R}_{\geq 0} \to \mathbb{R}_{\geq 0}$, and $T_n : [-1, 1] \to [-1, 1]$.

Exercise 5.6. Complete the following definition for polynomials represented as a plain list of coefficients:

```
instance Num a ⇒ Num [a] where
  (+) = addP
  (*) = mulP
    -- ... TODO
addP :: Num a ⇒ [a] → [a] → [a]
addP = zipWith' (+)
mulP :: Num a ⇒ [a] → [a] → [a]
mulP =    -- TODO
```

Note that *zipWith'* is almost, but not quite, the definition of *zipWith* from the standard Haskell prelude.

Exercise 5.7. What are the ring operations on *Poly'* A where

$$Poly' \; A = \{a : \mathbb{N} \to A \mid \{\text{- } a \text{ has a finite number of non-zero values -}\} \; \}$$

Exercise 5.8. Prove the *degree* law

$$\forall x, y. \; degree \; (x \; 'op' \; y) = degree \; x + degree \; y$$

for polynomials.

Exercise 5.9. Is *degree* a homomorphism from the monoid *(Poly a, 1, *)* to the monoid *(Maybe* $\mathbb{N}$*, Just 0, opMaybe)*? Check all the *Monoid* and homomorphism properties in the claim.

Exercise 5.10. The helper function *polyMap* :: $(a \to b) \to (Poly \; a \to Poly \; b)$ that was used in the implementation of *negPoly* is a close relative of the usual *map* :: $(a \to b) \to ([a] \to [b])$. Both these are members of a type class called *Functor*:

```
class Functor f where
  fmap :: (a → b) → (f a → f b)
```

Implement an instance of *Functor* for *Maybe* and *ComplexSyn* from Chapter 1. Is *fmap f* a homomorphism?

Exercise 5.11. Can the function *maximum* :: $[\mathbb{Z}] \to \mathbb{Z}$ be defined as a homo-morphism?

Solution: Some information is lacking, but we assume a monoid homomor-phism is requested. We write *maxi* instead of *maximum* below to keep it short. The source type (lists) is a monoid with $op = (+\!\!+)$ and $unit = [\,]$ and we are looking for a monoid structure on $\mathbb{Z}$. The homomorphism conditions are then *maxi* $(xs +\!\!+ ys) = op$ (*maxi xs*) (*maxi ys*) and *maxi* $[\,] = unit$, for some operation $op : \mathbb{Z} \to \mathbb{Z} \to \mathbb{Z}$ and a constant $unit : \mathbb{Z}$, forming a monoid.

Because of what maximum does, we must pick $op = max$. The *unit* must act like an identity for *max*: *max unit x* = *x*. This is possible only if $unit \leq x$ for every x. But, this is not possible: there is no lower bound in $\mathbb{Z}$. Thus, *maxi* is *not* a (monoid) homomorphism from $[\mathbb{Z}]$ to $\mathbb{Z}$. (If we create another type $\mathbb{Z}'$ with $-\infty$ added to $\mathbb{Z}$ we could define another *maxi'* : $[\mathbb{Z}] \to \mathbb{Z}'$).

Chapter 6

Taylor and Maclaurin series

In this chapter we make heavy use of concepts from Chapter 4 and thus we urge readers to verify their understanding of Sections 4.7.1 and 4.9 in case they might have skipped it. We have seen in particular that we can give a numeric (*Field*, etc.) structure not only to functions, but also to pairs of functions and their derivatives (*Field* $x \Rightarrow$ *Field* $(a \rightarrow x, a \rightarrow x)$). But why stop there? Why not compute a (lazy) list (also called a stream) of a function together with its derivative, second derivative, etc.:

$$[f, f', f'', \ldots] :: [a \rightarrow a]$$

The above then represents the evaluation of a 1-variable expression as a function, and all its derivatives. We can write this evaluation as an explicit function:

$$evalAll :: Transcendental\ a \Rightarrow FunExp \rightarrow [a \rightarrow a]$$
$$evalAll\ e = (evalFunExp\ e) : evalAll\ (derive\ e)$$

However *evalAll* is a non-compositional way of computing this stream. We will now proceed to define a specification of *evalAll* (as a homomorphism), and then derive a compositional implementation. Along the way we will continue to build insight about such streams of derivatives.

Notice that if we look at our stream of derivatives,

$$[f, f', f'', \ldots] = evalAll\ e$$

then the tail is also such a stream, but starting from f':

$[f',f'',\ldots] = evalAll\ (derive\ e)$

Thus $evalAll\ (derive\ e)$ == $tail\ (evalAll\ e)$ which can be written $evalAll \circ derive = tail \circ evalAll$. Thus $evalAll$ is a homomorphism from *derive* to *tail*, or in other words, we have $H_1(evalAll, derive, tail)$ (H_1 was defined in Exercise 4.18) — this is our specification for what follows.

We want to define the other numeric operations on streams of derivatives in such a way that *evalAll* is a homomorphism in each of them. For example, consider multiplication:

$$evalAll\ (e_1 :*: e_2) = evalAll\ e_1 * evalAll\ e_2$$

where the $(*)$ sign stands for the multiplication of derivative streams — an operation we are trying to determine, not the usual multiplication. We assume that we have already derived the definition of $(+)$ for these streams (it is *zipWithLonger* $(+)$, or just *zipWith* $(+)$ if we stick to infinite streams only).

We have the following derivation (writing *eval* for *evalFunExp* and *d* for *derive* in order to get a better overview):

> LHS
> $= \{-$ def. $-\}$
> $evalAll\ (e_1 :*: e_2)$
> $= \{-$ def. of *evalAll* $-\}$
> $eval\ (e_1 :*: e_2) : evalAll\ (d\ (e_1 :*: e_2))$
> $= \{-$ def. of *eval* for $(:*:)$ $-\}$
> $(eval\ e_1 * eval\ e_2) : evalAll\ (d\ (e_1 :*: e_2))$
> $= \{-$ def. of *derive* for $(:*:)$ $-\}$
> $(eval\ e_1 * eval\ e_2) : evalAll\ ((d\ e_1 :*: e_2) :+: (e_1 * d\ e_2))$
> $= \{-$ we assume $H_2(evalAll, (:+:), (+))$ $-\}$
> $(eval\ e_1 * eval\ e_2) : (evalAll\ (d\ e_1 :*: e_2) + evalAll\ (e_1 :*: d\ e_2))$

Similarly, starting from the other end we get

> $evalAll\ e_1 * evalAll\ e_2$
> $= \{-$ def. of *evalAll*, twice $-\}$
> $(eval\ e_1 : evalAll\ (d\ e_1)) * (eval\ e_2 : evalAll\ (d\ e_2))$

Now, to see the pattern it is useful to give simpler names to some common subexpressions: let $a = eval\ e_1$ and $b = eval\ e_2$.

> $(a * b) : (evalAll\ (d\ e_1 :*: e_2) + evalAll\ (e_1 * d\ e_2))$
> $=?$
> $(a : evalAll\ (d\ e_1)) * (b : evalAll\ (d\ e_2))$

Now we can solve part of the problem by defining $(*)$ as

$$(a : as) * (b : bs) = (a * b) : help\ a\ b\ as\ bs$$

The remaining part is then

$$evalAll\ (d\ e_1 :*: e_2) + evalAll\ (e_1 :*: d\ e_2)$$
$$=?$$
$$help\ a\ b\ (evalAll\ (d\ e_1))\ (evalAll\ (d\ e_2))$$

We now have two terms of the same form as we started out from: calls of *evalAll* on the constructor $(:*:)$. If we assume the homomorphism condition holds for these two calls we can rewrite *evalAll* $(d\ e_1 :*: e_2)$ to *evalAll* $(d\ e_1) *$ *evalAll* e_2 and similarly for the second term. (For a formal proof we also need to check that this assumption can be discharged.)

We also have *evalAll* $\circ\ d = tail \circ evalAll$ which leads to:

$$tail\ (evalAll\ e_1) * evalAll\ e_2 + evalAll\ e_1 * tail\ (evalAll\ e_2)$$
$$=?$$
$$help\ a\ b\ (tail\ (evalAll\ e_1))\ (tail\ (evalAll\ e_2))$$

Finally we rename common subexpressions: let $a : as = evalAll\ e_1$ and $b : bs = evalAll\ e_2$.

$$tail\ (a : as) * (b : bs) + (a : as) * tail\ (b : bs)$$
$$=?$$
$$help\ a\ b\ (tail\ (a : as))\ (tail\ (b : bs))$$

This equality is clearly satisfied if we define *help* as follows:

$$help\ a\ b\ as\ bs = as * (b : bs) + (a : as) * bs$$

Thus, we can eliminate *help* to arrive at a definition for multiplication[1]:

$$mulStream :: Ring\ a \Rightarrow Stream\ a \to Stream\ a \to Stream\ a$$
$$mulStream\ []\quad\quad \_\quad\ = []$$
$$mulStream\ \_\quad\quad []\quad = []$$
$$mulStream\ (a : as)\ (b : bs) = (a * b) : (as * (b : bs) + (a : as) * bs)$$

As in the case of pairs, we find that we do not need any properties of functions, other than their *Ring* structure, so the definitions apply to any infinite list of *Ring* values which we call a *Stream*:

---

[1]This expression is reminiscent of polynomial multiplication (Section 5.1), but it is different from it because here each element is implicitly divided by a factorial, as we shall see below. Hence we compute several terms many times here, and sum them together.

type *Stream a* $= [a]$
instance *Additive a* $\Rightarrow$ *Additive* (*Stream a*) **where**
  zero $=$ *repeat zero*
  $(+) =$ *addStream*
instance *AddGroup a* $\Rightarrow$ *AddGroup* (*Stream a*) **where**
  negate $=$ *negStream*
instance *Ring a* $\Rightarrow$ *Multiplicative* (*Stream a*) **where**
  one $=$ *one* : *zero*
  $(*) =$ *mulStream*

addStream :: *Additive a* $\Rightarrow$ *Stream a* $\rightarrow$ *Stream a* $\rightarrow$ *Stream a*
addStream $=$ *zipWithLonger* $(+)$

negStream :: *AddGroup a* $\Rightarrow$ *Stream a* $\rightarrow$ *Stream a*
negStream $=$ *map negate*

Exercise 6.1. Complete the *Stream* instance declarations for *MulGroup* and *Transcendental*.

Note that it may make more sense to declare a **newtype** for *Stream a* instead of using $[a]$, for at least two reasons. First, because the type $[a]$ also contains finite lists, but we use *Stream* here to represent only the infinite lists. Second, because there are competing possibilities for *Ring* instances for infinite lists, for example applying all the operations indexwise.[2] We used just a type synonym here to avoid cluttering the definitions with the **newtype** constructors.

Exercise 6.2. Write a general derivative computation: *drv k f x* $=$ the *k*th derivative of *f* at *x*.

6.1   Taylor series

We have arrived at the instances for *Stream* $(a \rightarrow a)$ by reasoning about lists of functions. But everywhere we needed to manipulate functions, we ended up using their numerical structure directly (assuming instances such as *Ring* $(x \rightarrow a)$, rather than treating them at functions). So, the *Stream* instances hold for *any* numeric type *a*. Effectively, we have implicitly used the *apply* homomorphism, as we did in Section 4.6.1. So we can view *Stream a* as series of higher-order derivatives taken at the same point *a*:

$$[f~(a), f'~(a), f''~(a), \ldots]$$

---

[2]These can be obtained applying the homomorphism between $[a]$ and $\mathbb{N} \rightarrow a$ to the *Ring* instances of $(x \rightarrow a)$

Assume now that f is a power series of coefficients (a_i):

$$f = eval\ [a_0, a_1, \ldots, a_n, \ldots]$$

We derive:

$$
\begin{aligned}
f\ 0 &= a_0 \\
f' &= eval\ (deriv\ [a_0, a_1, \ldots, a_n, \ldots]) \\
&= eval\ ([1 * a_1, 2 * a_2, 3 * a_3, \ldots, n * a_n, \ldots]) \\
\Rightarrow \\
f'\ 0 &= 1 * a_1 \\
f'' &= eval\ (deriv\ [1 * a_1, 2 * a_2, \ldots, n * a_n, \ldots]) \\
&= eval\ ([2 * 1 * a_2, 3 * 2 * a_3, \ldots, n * (n - 1) * a_n, \ldots]) \\
\Rightarrow \\
f''\ 0 &= 2 * 1 * a_2
\end{aligned}
$$

In general:

$$f^{(k)} 0 = fact\ k * a_k$$

Therefore

$$f = eval\ [f\ 0, f'\ 0, f''\ 0\ /\ 2, \ldots, f^{(n)} 0\ /\ (fact\ n), \ldots]$$

That is, there is a simple mapping between the representation of f as a power series (the coefficients a_k), and the value of all derivatives of f at 0.

The power series represented by $[f\ 0, f'\ 0, f''\ 0\ /\ 2, \ldots, f^{(n)} 0\ /\ (fact\ n), \ldots]$ is called the Taylor series centred at 0, or the Maclaurin series.

type *Taylor a = Stream a*

We can perform the above mapping (between a power series and its Maclaurin series) efficiently as follows:

toMaclaurin :: Ring a ⇒ PowerSeries a → Taylor a
toMaclaurin (Poly as) = zipWith (*) as factorials

fromMaclaurin :: Field a ⇒ Taylor a → PowerSeries a
fromMaclaurin as = Poly (zipWith (/) as factorials)

using a list of all factorials (starting from 0):

factorials :: Ring a ⇒ [a]
factorials = factorialsFrom 0 1

factorialsFrom :: Ring a ⇒ a → a → [a]
*factorialsFrom n factn = factn : factorialsFrom (n + 1) (factn * (n + 1))*

Remember that $x = Poly\ [0,1]$:

> $ex3, ex4 :: (Eq\ a, Field\ a) \Rightarrow Taylor\ a$
> $ex3 = toMaclaurin\ (x^{\wedge}3 + two * x)$
> $ex4 = toMaclaurin\ sinx$

This means that the *Taylor* type, interpreted as a Maclaurin series, can work as an alternative representation for power series (and in certain cases it can be a better choice computationally).

Regardless, we can see *toMaclaurin* as a way to compute all the derivatives at 0 for all functions f constructed with the grammar of *FunExp*. That is because, as we have seen, we can represent all of them by power series!

What if we want the value of the derivatives at some other point a (different from zero)? We then need the power series of the "shifted" function g:

$$g\ x = f\ (a + x) \Leftrightarrow g = f \circ (a+)$$

If we can represent g as a power series, say $[b_0, b_1, \ldots]$, then we have

$$g^{(k)}0 = fact\ k * b_k = f^{(k)}a$$

In particular, we would have

$$f\ x = g\ (x - a) = \sum b_n * (x - a)^n$$

which is called the Taylor expansion of f at a.

Example: We have that $idx = Poly\ [0,1]$, thus giving us indeed the values

> $[id\ 0, id'\ 0, id''\ 0, \ldots]$

In order to compute the values of

> $[id\ a, id'\ a, id''\ a, \ldots]$

for $a \neq 0$, we compute

> $ida\ a = toMaclaurin\ (evalP\ (X :+: Const\ a))$

More generally, if we want to compute the derivatives of a function f constructed with *FunExp* grammar, at a point a, we can use the power series of $g\ x = f\ (x + a)$ (we additionally restrict ourselves to the first 10 derivatives):

> $d\ f\ a = take\ 10\ (toMaclaurin\ (evalP\ (f\ (X :+: Const\ a))))$

Use, for example, our $f\ x = sin\ x + 2 * x$ from Section 4.9. As before, we can use power series directly to construct the input:

> $dP\ f\ a = toMaclaurin\ (f\ (idx + Poly\ [a]))$

6.2 Derivatives and Integrals for Maclaurin series

Since the Maclaurin series represents $[f\ 0, f'\ 0, f''\ 0, \ldots]$, the tail of the list is equivalent to the derivative of f. To prove that fact, one can substitute f by f' in the above. Another way to see it is to remark that we started with the equation *evalAll* ∘ *derive* = *tail* ∘ *evalAll*; but now our input representation is *already* a Maclaurin series so *evalAll* = *id*, and in turn *derive* = *tail*. In sum:

> *head f* = *eval f* 0 -- value of f at 0
> *tail f* = *deriv f* -- derivative of f

Additionally, integration can be defined simply as the list constructor "cons" with the first argument being the value of f at 0:

> *integT* :: $a \rightarrow$ *Taylor a* $\rightarrow$ *Taylor a*
> *integT* = (:)

Given that we have an infinite list, we have $f = head\ f : tail\ f$. Let's see what this law means in terms of calculus:

> f == *head f* : *tail f*
> f == (:)   (*head f*) (*tail f*)
>   == *integT* (*head f*) (*tail f*)
>   == *integT* f_0 (*deriv f*)

or, in traditional notation:

$$f(x) = f(0) + \int_0^x f'(t)\mathrm{d}t$$

which is the fundamental theorem of calculus [Adams and Essex, 2010, Sect. 5.5]. In sum, we find that our definition of *integT* is exactly that needed to comply with calculus laws.

6.3 Integral for Formal Power series

In Section 5.6 we found a definition of derivatives for formal power series (which we can also divide, as well as add and multiply):

> *deriv* :: *Ring a* $\Rightarrow$ *PowerSeries a* $\rightarrow$ *PowerSeries a*
> *deriv* (*Poly* [])   = *Poly* []
> *deriv* (*Poly* (_ : *as*)) = *Poly* (*zipWith* (∗) *oneUp as*)

$$oneUp :: Ring\ a \Rightarrow [a]$$
$$oneUp = countUp\ one$$

$$countUp :: Ring\ a \Rightarrow a \rightarrow [a]$$
$$countUp = iterate\ (one+)$$

With our insight regarding Taylor series, we can now see that *toMaclaurin* ∘ *deriv* == *tail* ∘ *toMaclaurin*. We can apply the same recipe to obtain integration for power series: *toMaclaurin* ∘ *integ* a_0 == *integT* a_0 ∘ *toMaclaurin*. If we see this as the specification of *integ* and carry out the calculation we arrive at:

$$integ\ :: Field\ a \Rightarrow a \rightarrow PowerSeries\ a \rightarrow PowerSeries\ a$$
$$integ\ a_0\ (Poly\ as) = Poly\ (integL\ a_0\ as)$$

$$integL :: Field\ a \Rightarrow a \rightarrow [a] \rightarrow [a]$$
$$integL\ c\ cs = c : zipWith\ (/)\ cs\ oneUp$$

Remember that a_0 is the constant that we need due to indefinite integration.

These operations work on the type *PowerSeries a* which we can see as the syntax of power series, often called "formal power series". The intended semantics of a formal power series *a* is, as we saw in Chapter 5, an infinite sum

$$eval\ a : \mathbb{R} \rightarrow \mathbb{R}$$
$$eval\ a = \lambda x \rightarrow lim\ s\ \textbf{where}\ s\ n = \sum_{i=0}^{n} a_i * x^i$$

For any *n*, the prefix sum, *s n*, is finite and it is easy to see that the derivative and integration operations are well defined. When we take the limit, however, the sum may fail to converge for certain values of *x*. Fortunately, we can often ignore that, because seen as operations from syntax to syntax, all the operations are well defined, irrespective of convergence.

If the power series involved do converge, then *eval* is a morphism between the formal structure and that of the functions represented:

$$eval\ as + eval\ bs\ \ = eval\ (as + bs)\quad \text{-- } H_2(eval, (+), (+))$$
$$eval\ as * eval\ bs\ \ = eval\ (as * bs)\quad \text{-- } H_2(eval, (*), (*))$$
$$eval\ (derive\ as)\ \ = D\ (eval\ as)\qquad \text{-- } H_1(eval, derive, D)$$
$$eval\ (integ\ c\ as)\ x = c + \int_0^x (eval\ as\ t)\ dt$$

6.4   Simple differential equations

Many first-order differential equations have the structure

$$f' x = g f x, \quad f 0 = f_0$$

i.e., they are defined in terms of the higher-order function g and initial value f_0. The fundamental theorem of calculus gives us

$$f x = f_0 + \int_0^x (g f\, t)\, dt$$

If $f = eval\ as$

$$eval\ as\ x = f_0 + \int_0^x (g\ (eval\ as)\ t)\, dt$$

Assuming that g is a polymorphic function defined both for the syntax ($PS\ \mathbb{R}$) and the semantics ($\mathbb{R} \to \mathbb{R}$), and that

$$\forall\ as.\ eval\ (g_{syn}\ as) == g_{sem}\ (eval\ as)$$

or simply $H_1(eval, g, g)$. (This particular use of H_1 is read "g commutes with *eval*".) Then we can move *eval* outwards step by step:

$$eval\ as\ x = f_0 + \int_0^x (eval\ (g\ as)\ t)\, dt$$
$$\Leftrightarrow eval\ as\ x = eval\ (integ\ f_0\ (g\ as))\ x$$
$$\Leftarrow as = integ\ f_0\ (g\ as)$$

Finally, we have arrived at an equation expressed in only syntactic operations, which is implementable in Haskell (for a reasonable g).

Which functions g commute with *eval*? All functions built from methods in *Ring*, *Field*, *Transcendental*, by construction; additionally, as above, *deriv* and *integ*. Therefore, we can implement a general solver for these simple equations:

type $PS\ a = PowerSeries\ a$
$solve :: Field\ a \Rightarrow a \to (PS\ a \to PS\ a) \to PS\ a$
$solve\ f_0\ g = f$ $\quad$ -- solves $f' = g f, f\ 0 = f_0$
$\quad$ **where** $f = integ\ f_0\ (g f)$

On the face of it, the solution f appears not well defined, because its definition depends on itself. We come back to this point soon, but first we observe *solve* in action on simple instances of g, starting with *const* 1 and *id*:

$idx :: Field\ a \Rightarrow PS\ a$
$idx = solve\ 0\ (\backslash\_f \to 1)$ $\quad$ -- $f'(x) = 1, f(0) = 0$
$expx :: Field\ a \Rightarrow PS\ a$
$expx = solve\ 1\ (\lambda f \to f)$ $\quad$ -- $f'(x) = f(x), f(0) = 1$
$expf :: Field\ a \Rightarrow a \to a$
$expf = evalPS\ 100\ expx$

Exercise 6.3. Write *expx* as a recursive equation (inline *solve* in the definition).

The first solution, *idx* is just the polynomial $[0, 1]$ — i.e. just x in usual mathematical notation. We can easily check that its derivative is constantly 1 and its value at 0 is 0.

The second solution *expx* is a formal power series representing the exponential function. It is equal to its derivative and it starts at 1. The function *expf* is a good approximation of the semantics for small values of its argument — the following testing code shows that the maximum deviation in the interval from 0 to 1 is below $5 * 10^{-16}$ (very close to the precision of *Double*).

> *testExp* :: *Double*
> *testExp* = *maximum* (*map diff* $[0, 0.001 .. 1 :: Double]$)
>   **where** *diff* = *abs* ∘ (*expf* − *exp*)   -- using the function instance for *exp*
> *testExpUnits* :: *Double*
> *testExpUnits* = *testExp* / ϵ
>
> ϵ :: *Double*   -- one bit of *Double* precision
> ϵ = *last* (*takeWhile* ($\lambda x \rightarrow 1 + x \neq 1$) (*iterate* (/2) 1))

As an alternative to using *solve* we can use recursion directly. For example, we can define sine and cosine in terms of each other:

> *sinx*, *cosx* :: *Field a* $\Rightarrow$ *PS a*
> *sinx* = *integ* 0 *cosx*
> *cosx* = *integ* 1 (−*sinx*)
>
> *sinf*, *cosf* :: *Field a* $\Rightarrow$ *a* $\rightarrow$ *a*
> *sinf* = *evalPS* 100 *sinx*
> *cosf* = *evalPS* 100 *cosx*

Exercise 6.4. Write the differential equations characterising sine and cosine, using usual mathematical notation.

The reason that these definitions produce an output instead of entering an infinite loop is that Haskell is a lazy language: *integ* can immediately return the first element of the stream before requesting any information about its second input. It is instructive to mimic part of what the lazy evaluation machinery is doing by hand, as follows. We know that both *sinx* and *cosx* are streams, thus we can start by filling in just the very top level structure:

> *sx* = *sh* : *st*
> *cx* = *ch* : *ct*

where *sh* & *ch* are the heads and *st* & *ct* are the tails of the two streams. Then we notice that *integ* fills in the constant as the head, and we can progress to:

$$sx = 0 : st$$
$$cx = 1 : ct$$

At this stage we only know the constant term of each power series, but that is enough for the next step: the head of *st* is $\frac{1}{1}$ and the head of *ct* is $\frac{-0}{1}$:

$$sx = 0 : 1 : \_$$
$$cx = 1 : -0 : \_$$

As we move on, we can always compute the next element of one series by the previous element of the other series (divided by n, for *cx* negated).

$$sx, cx :: [Double]$$
$$sx = 0 : 1 : -0 : \tfrac{-1}{6} : error \text{ "TODO"}$$
$$cx = 1 : -0 : \tfrac{-1}{2} : 0 : error \text{ "TODO"}$$

6.5 Exponentials and trigonometric functions

We have now shown how to compute the power series representations of the functions *exp*, *sin*, and *cos*. We have also implemented all the *Field* class operations on power series. The next step would be to compute the *Transcendental* class operations directly on the power series representation. For example, can we compute *expPS*?

Specification:

$$eval \ (expPS \ as) = exp \ (eval \ as)$$

Note that *expx* : *PS a* is a power series in itself, but *expPS* : *PS a* → *PS a* is a "power series transformer".

Differentiating both sides of the specification, we obtain

$$D \ (eval \ (expPS \ as)) = exp \ (eval \ as) * D \ (eval \ as)$$
$$\Leftrightarrow \{\text{- } eval \text{ homomorphism -}\}$$
$$eval \ (deriv \ (expPS \ as)) = eval \ (expPS \ as * deriv \ as)$$
$$\Leftarrow$$
$$deriv \ (expPS \ as) = expPS \ as * deriv \ as$$

Now we have reached the form of a differential equation for *expPS as*, and we know how to solve them by integration, given the initial condition. Using *eval (expPS as) 0 == exp (eval as 0) == exp (head as)*, we obtain

$$expPS\ as = integ\ (exp\ (head\ as))\ (expPS\ as * deriv\ as)$$

Note: we cannot use *solve* here, because the *g* function uses both *expPS as* and *as* (it "looks inside" its argument).

In the same style we can fill in the *Transcendental* instance declaration for *PowerSeries*:

instance (Eq a, Transcendental a) ⇒ Transcendental (PowerSeries a) **where**
   π   = Poly [π]
   exp = expPS
   sin = sinPS
   cos = cosPS
expPS, sinPS, cosPS :: (Eq a, Transcendental a) ⇒ PS a → PS a
expPS as = integ (exp (val as)) (expPS as   * deriv as)
sinPS as = integ (sin (val as)) (cosPS as   * deriv as)
cosPS as = integ (cos (val as)) (−sinPS as * deriv as)
val :: Additive a ⇒ PS a → a
val (Poly (a: _)) = a
val _   = zero

Note that we have defined *val* as a more robust version of *head* when it comes to computing the value of the function at zero.

In fact, we can now implement *all* the operations needed for evaluating *FunExp* functions as power series!

evalP :: (Eq r, Transcendental r) ⇒ FunExp → PS r
evalP (Const x) = Poly [fromRational (toRational x)]
evalP (e₁ :+: e₂) = evalP e₁ + evalP e₂
evalP (e₁ :*: e₂) = evalP e₁ * evalP e₂
evalP X   = idx
evalP (Negate e) = negate (evalP e)
evalP (Recip e) = recip (evalP e)
evalP (Exp e)   = exp   (evalP e)
evalP (Sin e)   = sin   (evalP e)
evalP (Cos e)   = cos   (evalP e)

6.6   Associated code

Here we collect in one place the definitions of *eval* for *FunExp*, syntactic derivative, syntactic instance declarations for *FunExp*, and numeric instances for
pairs $(f\ a, f'\ a)$.

6.6.1   Full definition of *evalFunExp*

Note the use of "lifted $+$", "lifted $*$", "lifted *exp*", etc. on the right-hand sides
of *evalFunExp*.

$$
\begin{array}{ll}
\textit{evalFunExp} :: \textit{Transcendental } a \Rightarrow \textit{FunExp} \rightarrow a \rightarrow a \\
\textit{evalFunExp } (\textit{Const } \alpha) & = \textit{const } (\textit{fromRational } (\textit{toRational } \alpha)) \\
\textit{evalFunExp } X & = \textit{id} \\
\textit{evalFunExp } (e_1 \mathbin{:+:} e_2) & = \textit{evalFunExp } e_1 + \textit{evalFunExp } e_2 \\
\textit{evalFunExp } (e_1 \mathbin{:*:} e_2) & = \textit{evalFunExp } e_1 * \textit{evalFunExp } e_2 \\
\textit{evalFunExp } (\textit{Exp } e) & = \textit{exp} \quad (\textit{evalFunExp } e) \\
\textit{evalFunExp } (\textit{Sin } e) & = \textit{sin} \quad (\textit{evalFunExp } e) \\
\textit{evalFunExp } (\textit{Cos } e) & = \textit{cos} \quad (\textit{evalFunExp } e) \\
\textit{evalFunExp } (\textit{Recip } e) & = \textit{recip} \quad (\textit{evalFunExp } e) \\
\textit{evalFunExp } (\textit{Negate } e) & = \textit{negate } (\textit{evalFunExp } e)
\end{array}
$$

6.6.2   Syntactic derivative: *derive* : *FunExp* $\rightarrow$ *FunExp*

Note that this syntactic computation of the derivative does not perform any
kind of algebraic simplification of the result. So, for example:

$$
\textit{derive } (X \mathbin{:*:} X) == (\textit{Const } 1.0 \mathbin{:*:} X) \mathbin{:+:} (X \mathbin{:*:} \textit{Const } 1.0)
$$

See Exercise 4.28 for a suggestion to improve on this.

$$
\begin{array}{ll}
\textit{derive } (\textit{Const } \_) & = \textit{Const } 0 \\
\textit{derive } X & = \textit{Const } 1 \\
\textit{derive } (e_1 \mathbin{:+:} e_2) & = \textit{derive } e_1 \mathbin{:+:} \textit{derive } e_2 \\
\textit{derive } (e_1 \mathbin{:*:} e_2) & = (\textit{derive } e_1 \mathbin{:*:} e_2) \mathbin{:+:} (e_1 \mathbin{:*:} \textit{derive } e_2) \\
\textit{derive } (\textit{Recip } e) & = \textbf{let } re = \textit{Recip } e \textbf{ in } \textit{Negate } (re \mathbin{:*:} re) \mathbin{:*:} \textit{derive } e \\
\textit{derive } (\textit{Negate } e) & = \textit{Negate } (\textit{derive } e) \\
\textit{derive } (\textit{Exp } e) & = \textit{Exp } e \mathbin{:*:} \textit{derive } e \\
\textit{derive } (\textit{Sin } e) & = \textit{Cos } e \mathbin{:*:} \textit{derive } e \\
\textit{derive } (\textit{Cos } e) & = \textit{Const } (-1) \mathbin{:*:} \textit{Sin } e \mathbin{:*:} \textit{derive } e
\end{array}
$$

6.6.3   Numeric instances for *FunExp*

```
instance Additive FunExp where
  (+) = (:+:)
  zero = Const 0
instance AddGroup FunExp where
  negate x = Const (−1) * x
instance Multiplicative FunExp where
  (*) = (:*:)
  one = Const 1
instance MulGroup FunExp where
  recip = Recip
instance Transcendental FunExp where
  π   = Const π
  exp = Exp
  sin = Sin
  cos = Cos
```

6.6.4   Numeric instances for *Dup*

For working with a value of a function and its derivative at a point.

```
instance Additive a ⇒ Additive (a, a) where
  (f, f′) + (g, g′) = (f + g, f′ + g′)
  zero            = (zero,  zero)
instance AddGroup a ⇒ AddGroup (a, a) where
  negate (f, f′)   = (negate f, negate f′)
instance Ring a ⇒ Multiplicative (a, a) where
  (f, f′) * (g, g′) = (f * g, f′ * g + f * g′)
  one             = (one, zero)
instance Field a ⇒ MulGroup (a, a) where
  (f, f′) / (g, g′) = (f / g, (f′ * g − g′ * f) / (g * g))
instance Transcendental a ⇒ Transcendental (a, a) where
  π = (π, zero)
  exp (f, f′)     = (exp f, (exp f) * f′)
  sin (f, f′)     = (sin f, cos f * f′)
  cos (f, f′)     = (cos f, − (sin f) * f′)
```

6.7 Exercises

Exercise 6.5. As shown in Section 4.9, we can find expressions $e :: FunExp$ such that $eval\ e = f$ automatically using the assignment $e = f\ Id$. This is possible thanks to the *Ring, Transcendental*, etc. instances of *FunExp*. Use this method to find *FunExp* representations of the functions below, and show step by step how the application of the function to *Id* is evaluated in each case.

1. $f_1\ x = x\char`^2 + 4$

2. $f_2\ x = 7 * exp\ (2 + 3 * x)$

3. $f_3\ x = 1\ /\ (sin\ x + cos\ x)$

Exercise 6.6. For each of the expressions $e :: FunExp$ you found in Exercise 6.5, use *derive* to find an expression $e' :: FunExp$ representing the derivative of the expression, and verify that e' is indeed the derivative of e.

Exercise 6.7. At the start of this chapter, we saw three different ways of computing the value of the derivative of a function at a given point:

1. Using *FunExp*

2. Using *FD*

3. Using pairs

Try using each of these methods to find the values of $f_1'\ 2$, $f_2'\ 2$, and $f_3'\ 2$, i.e. the derivatives of each of the functions in Exercise 6.5, evaluated at the point 2. You can verify that the result is correct by comparing it with the expressions e_1', e_2' and e_3' that you found in 6.6.

Exercise 6.8. The exponential function $exp\ t = e^t$ has the property that $\int e^t dt = e^t + C$. Use this fact to express the functions below as *PowerSeries* using *integ*. *Hint: the definitions will be recursive.*

1. $\lambda t \rightarrow exp\ t$

2. $\lambda t \rightarrow exp\ (3 * t)$

3. $\lambda t \rightarrow 3 * exp\ (2 * t)$

Exercise 6.9. In the chapter, we saw that a representation $expx :: PowerSeries$ of the exponential function can be implemented using *solve* as $expx = solve\ 1\ id$. Use the same method to implement power series representations of the following functions:

1. $\lambda t \to exp\,(3 * t)$

2. $\lambda t \to 3 * exp\,(2 * t)$

Exercise 6.10.

Implement *sinx* and *cosx* using *solve* instead of a recursive definition.

Exercise 6.11. Consider the following differential equation:

$$f''\,t + f'\,t - 2 * f\,t = e^{3*t}, \quad f\,0 = 1, \quad f'\,0 = 2$$

We will solve this equation assuming that f can be expressed by a power series *fs*, and finding the three first coefficients of *fs*.

1. Implement *expx3* :: *PowerSeries* $\mathbb{R}$, a power series representation of e^{3*t}

2. Find an expression for *fs''*, the second derivative of *fs*, in terms of *expx3*, *fs'*, and *fs*.

3. Find an expression for *fs'* in terms of *fs''*, using *integ*.

4. Find an expression for *fs* in terms of *fs'*, using *integ*.

5. Use *takePoly* to find the first three coefficients of *fs*. You can check that your solution is correct using a tool such as MATLAB or WolframAlpha, by first finding an expression for $f\,t$, and then getting the Taylor series expansion for that expression.

Exercise 6.12. *From exam 2016-03-15*

Consider the following differential equation:

$$f''\,t - 2 * f'\,t + f\,t = e^{2*t}, \quad f\,0 = 2, \quad f'\,0 = 3$$

Solve the equation assuming that f can be expressed by a power series *fs*, that is, use *deriv* and *integ* to compute *fs*. What are the first three coefficients of *fs*?

Exercise 6.13. *From exam 2016-08-23*

Consider the following differential equation:

$$f''\,t - 5 * f'\,t + 6 * f\,t = e^{t}, \quad f\,0 = 1, \quad f'\,0 = 4$$

Solve the equation assuming that f can be expressed by a power series *fs*, that is, use *deriv* and *integ* to compute *fs*. What are the first three coefficients of *fs*?

Exercise 6.14. *From exam 2016-Practice*

Consider the following differential equation:

$$f'' \, t - 2 * f' \, t + f \, t - 2 = 3 * e^{2*t}, \quad f \, 0 = 5, \quad f' \, 0 = 6$$

Solve the equation assuming that f can be expressed by a power series fs, that is, use *deriv* and *integ* to compute fs. What are the first three coefficients of fs?

Exercise 6.15. *From exam 2017-03-14*

Consider the following differential equation:

$$f'' \, t + 4 * f \, t = 6 * \cos t, \quad f \, 0 = 0, \quad f' \, 0 = 0$$

Solve the equation assuming that f can be expressed by a power series fs, that is, use *integ* and the differential equation to express the relation between fs, fs', fs'', and *rhs* where *rhs* is the power series representation of $(6*) \circ cos$. What are the first four coefficients of fs?

Exercise 6.16. *From exam 2017-08-22*

Consider the following differential equation:

$$f'' \, t - 3\sqrt{2} * f' \, t + 4 * f \, t = 0, \quad f \, 0 = 2, \quad f' \, 0 = 3\sqrt{2}$$

Solve the equation assuming that f can be expressed by a power series fs, that is, use *integ* and the differential equation to express the relation between fs, fs', and fs''. What are the first three coefficients of fs?

Chapter 7

Elements of Linear Algebra

Often, especially in engineering textbooks, one encounters the following definition: a vector is an $n + 1$-tuple of real or complex numbers, arranged as a column:

$$v = \begin{bmatrix} v_0 \\ \vdots \\ v_n \end{bmatrix}$$

Other times, this is supplemented by the definition of a row vector:

$$v = \begin{bmatrix} v_0 & \cdots & v_n \end{bmatrix}$$

The v_is are real or complex numbers, or, more generally, elements of a *field* (See Section 4.1.1 for the definition of a field).

Vectors and their spaces   However, following our theme, we will first characterise vectors algebraically. From this perspective a *vector space* is an algebraic structure that captures a set of vectors, with zero, a commutative addition, and scaling by a set of scalars (i.e., elements of the field). In terms of type classes, we can characterise this structure as follows:

> **class** *(Field s, AddGroup v)* $\Rightarrow$ *VectorSpace v s* **where**
> $(\triangleleft) :: s \rightarrow v \rightarrow v$

The class declaration is short, but can need some unpacking. First, the type s for the scalars, is required to be a field, which basically means we have $(+)$, $(-)$, $(*)$, and $(/)$ available as operations on values of type s, the scale factors. Then, the type v needs to be an additive group, thus supporting a zero vector, $(+)$ and $(-)$ on vectors. These operations are all required before we are allowed to declare a *VectorSpace* instance, due to the constraint $(Field\ s, AddGroup\ v)$. Finally, the new operator $(\lhd)$, called "scale" or scalar-vector-multiplication, takes a scale factor and a vector to a suitably resized vector: $2 \lhd v$ is twice v, while $(-1) \lhd v$ has the same length as v but points in the opposite direction, etc.

Laws   Additionally, every vector space must satisfy the following laws:

1. Vector scaling $((s\lhd) :: v \to v)$ is a homomorphism over (from and to) the additive group structure of v. Thus for all vectors a and b we have:

$$
\begin{aligned}
s \lhd (a+b) \quad &= s \lhd a + s \lhd b \\
s \lhd zero \quad &= zero \\
s \lhd (negate\ a) &= negate\ (s \lhd a)
\end{aligned}
$$

   This means that scaling can be "pushed inside" any sum or difference.

2. On the other side, $(\lhd a)$ is a homomorphism from the additive group structure of s to the group structure of v. Thus, for all scalars s and t we have:

$$
\begin{aligned}
(s+t) \quad \lhd a &= s \lhd a + t \lhd a \\
zero \quad \lhd a &= zero \\
negate\ s \lhd a &= negate\ (s \lhd a)
\end{aligned}
$$

   For the examples above this means that $2 \lhd v \mathrel{==} (1+1) \lhd v \mathrel{==} 1 \lhd v + 1 \lhd v$ and $(-1) \lhd v \mathrel{==} negate\ (1 \lhd v)$.

3. Finally $(\lhd)$ is a homomorphism from the multiplicative monoid of s to the monoid of endofunctions over v (see Section 4.1). Thus, for all scalars s and t we have:

$$
\begin{aligned}
(\lhd)\ one \quad &= id \\
(\lhd)\ (s*t) &= (\lhd)\ s \circ (\lhd)\ t
\end{aligned}
$$

   Applying the functions to the vector a everywhere gives the familiar form for these laws:

$$
\begin{aligned}
one \quad \lhd a &= id \qquad\qquad a = a \\
(s*t) \lhd a &= ((s\lhd) \circ (t\lhd))\ a = s \lhd (t \lhd a)
\end{aligned}
$$

   For the examples above this means that $2 \lhd v \mathrel{==} 1 \lhd v + 1 \lhd v \mathrel{==} v + v$ (or "twice v") and $(-1) \lhd v \mathrel{==} negate\ (1 \lhd v) \mathrel{==} negate\ v$ (or "v in the opposite direction").

Often, the above laws are not expressed in terms of homomorphisms, but rather as individual equations. This means that some of them are often omitted, because they are consequences of sets of other laws.

One-dimensional spaces We get the simplest instance declaration if we note that we can see scalars (like $\mathbb{R}$) as one-dimensional vectors with $s = v$:

> **instance** *Field s* $\Rightarrow$ *VectorSpace s s* **where** $(\lhd) = (*)$

Here (for once) the vectors have the same type as the scalars, which means that the scaling operation, which usually has an asymmetric type, now is just ordinary scalar multiplication $(*) :: s \to s \to s$. But for the rest of this chapter we will stick to the general case of n- (or infinite-) dimensional spaces.

Bases and representations An important consequence of the algebraic structure of vectors is that they can be expressed as a simple sort of combination of other special vectors. More precisely, we can *uniquely* represent any vector v in the space in terms of a fixed set of *basis* vectors $\{b_0, \ldots, b_n\}$. By definition, basis vectors cover the whole space:

$$\forall v.\ \exists s_0, \ldots, s_n.\ v == s_0 \lhd b_0 + \ldots + s_n \lhd b_n$$

They are also *linearly independent*:

$$(s_0 \lhd b_0 + \ldots + s_n \lhd b_n = 0) \Leftrightarrow (s_0 = \ldots = s_n = 0)$$

One can prove the uniqueness of representation as follows:

Proof. Assume two representations of v, given by s_i and t_i. The difference of those representations is given by $s_i - t_i$. But because they represent the same vector, their difference must be equal to the zero vector: $(s_0 - t_0) \lhd b_0 + \ldots + (s_n - t_n) \lhd b_n = 0$. By the basis being linearly independent, we find $s_i - t_i = 0$, and thus $s_i = t_i$. $\square$

Syntax for vectors According to our red thread, this representation (coefficients) is akin to the notion of syntax. But this is a case where the representation is *equivalent* to the algebraic definition: the evaluator is not only a homomorphism, but an isomorphism between the space of vectors and the list of coefficients. This equivalence is what justifies the definition of vectors as columns (or rows) of numbers.

Indeed, we can define:

$$v = \begin{bmatrix} v_0 \\ \vdots \\ v_n \end{bmatrix} = v_0 \triangleleft \begin{bmatrix} 1 \\ 0 \\ \vdots \\ 0 \end{bmatrix} + v_1 \triangleleft \begin{bmatrix} 0 \\ 1 \\ \vdots \\ 0 \end{bmatrix} + \cdots + v_n \triangleleft \begin{bmatrix} 0 \\ 0 \\ \vdots \\ 1 \end{bmatrix}$$

So, for our column vectors, we can define the operations as follows:

$$v + w = \begin{bmatrix} v_0 \\ \vdots \\ v_n \end{bmatrix} + \begin{bmatrix} w_0 \\ \vdots \\ w_n \end{bmatrix} = \begin{bmatrix} v_0 + w_0 \\ \vdots \\ v_n + w_n \end{bmatrix}$$

$$s \triangleleft v = \begin{bmatrix} s * v_0 \\ \vdots \\ s * v_n \end{bmatrix}$$

In the following we denote by

$$e_k = \begin{bmatrix} 0 \\ \vdots \\ 0 \\ 1 \\ 0 \\ \vdots \\ 0 \end{bmatrix} \quad \leftarrow \text{position } k$$

the canonical basis vectors, i.e. e_k is the vector that is everywhere zero except at position k, where it is one, so that $v == v_0 \triangleleft e_0 + \ldots + v_n \triangleleft e_n$. This formula maps the syntax (coefficients, v_i) to the semantics (a vector, v).

Exercise 7.1. Define a function which takes as input a vector v and a set of (non-canonical) basis vectors b_i and returns the coefficients of v in that basis.

7.1 Representing vectors as functions

In what follows we will systematically use the representation of vectors as a linear combination of basis vectors. There is a temptation to model the corresponding collection of coefficients as a list, or a tuple, but a more general (and conceptually simpler) way is to view them as a *function* from a set of indices G:

newtype *Vector s g* = *V* (*g* → *s*) **deriving** (*Additive, AddGroup*)

We define, right away, the notation *a* ! *i* for the coefficient of the canonical basis vector *e i*, as follows:

infix 9 !
(!) :: *Vector s g* → *g* → *s*
V f ! *i* = *f i*

We sometimes omit the constructor *V* and the indexing operator (!), thereby treating vectors as functions without the **newtype**. (We use the exclamation mark as an infix operator here as is common in programming, even though it is often used as postfix notation for factorial in mathematics texts.)

As discussed above, the *S* parameter in *Vector S* has to be a field ($\mathbb{R}$, or $\mathbb{C}$, or $\mathbb{Z}_p$[1], etc.) for values of type *Vector S G* to represent elements of a vector space.

The cardinality of *G*, which we sometimes denote *card G*, is the number of basis vectors, and thus the dimension of the vector space. Often *G* is finite, and in the examples so far we have used indices from $G = \{0, \ldots, n\}$. Thus the dimension of the space would be $n + 1$.

In Haskell finiteness of *G* can be captured by the conjunction of *Bounded* (there is a minimum and a maximum element in *G*) and *Enum*erable (there is a notion of enumeration from a given element of *G*) and *Eq*. Hence, the list of all elements of *G* can be extracted:

type *Finite g* = (*Bounded g, Enum g, Eq g*)
finiteDomain :: *Finite a* ⇒ [*a*]
finiteDomain = [*minBound .. maxBound*]

For our running example *finiteDomain* = [0, 1, 2, 3, 4, 5, 6].

Let us now define a *VectorSpace* instance for the *Vector* representation. This can only be done if *s* is a *Field*. Then, we must provide an associative and commutative addition operation. For *Vector*, it can is defined indexwise. Because indexwise addition is already our definition of addition for functions (*g* → *s*), from Section 3.5.3, we can simply reuse this definition. (Function addition demands that *s* is an instance of *AddGroup*, but it's fine since *s* is even a *Field*.) This is what the **deriving** clause amounts to in the definition of **newtype** *Vector*. The rest of the *AddGroup* structure, *zero* and *negate* is defined by the same means.

What about vector scaling, (◁)? Can we simply reuse the definition that we had for functions? No, because multiplication of vectors does not work pointwise. In fact, attempting to lift multiplication from the *Multiplicative* class

---

[1]The set $\mathbb{Z}_p$ is the set of integers modulo *p*. We let the reader lookup the appropriate notion of division for it.

would give a homogeneous multiplication operator $(*) :: v \rightarrow v \rightarrow v$, but such an operator is not part of the definition of vector spaces. Consequently, vector spaces are in general *not* rings.

Indeed, the scaling operator $(\lhd) :: s \rightarrow v \rightarrow v$, is inhomogeneous: the first argument is a scalar and the second one is a vector. For our representation it can be defined as follows:

> **instance** *Field s* $\Rightarrow$ *VectorSpace* (*Vector s g*) *s* **where** $(\lhd) = scaleV$
> *scaleV* :: *Multiplicative s* $\Rightarrow$ *s* $\rightarrow$ *Vector s g* $\rightarrow$ *Vector s g*
> *scaleV s* (*V a*) $= V$ ($\lambda i \rightarrow s * a\ i$)

Exercise 7.2. Show that *Vector s g* satisfies the laws of vector spaces.

The canonical basis for *Vector* are given by

> *e* :: (*Eq g, Ring s*) $\Rightarrow$ *g* $\rightarrow$ *Vector s g*
> *e i* $= V$ ($\lambda j \rightarrow i$ '*is*' j)

In linear algebra textbooks, the function *is* is often referred to as the Kronecker-delta function and *is i j* is written $\delta_{i,j}$.

> *is* :: (*Eq g, Ring s*) $\Rightarrow$ *g* $\rightarrow$ *g* $\rightarrow$ *s*
> *is i j* $=$ **if** *i* == *j* **then** *one* **else** *zero*

It is 1 if its arguments are equal and 0 otherwise. Thus *e i* has zeros everywhere, except at position *i* where it has a one.

We can see that, as expected, every $v : g \rightarrow s$ is a linear combination of vectors *e i* where the coefficient of the canonical basis vector *e i* is the scalar *v i*:

$$v == (v\ 0 \lhd e\ 0) + \ldots + (v\ n \lhd e\ n)$$

This property is called the *characterising equation* for vectors.

To be sure, every vector *v* is a linear combination of any collection of basis vectors. But when using *canonical* basis vectors, the coefficients come simply from applying *v* (seen as a function) to the possible indices. Because we will work with many such linear combinations we introduce a helper function *linComb* for the right-hand side of the characterising equation:

> *linComb* :: (*Finite g, VectorSpace v s*) $\Rightarrow$ (*g* $\rightarrow$ *s*) $\rightarrow$ (*g* $\rightarrow$ *v*) $\rightarrow$ *v*
> *linComb v e* $=$ *sum* (*map* ($\lambda j \rightarrow v\ j \lhd e\ j$) *finiteDomain*)

where you can think of *finiteDomain* as enumerating the indices $[0 .. n]$.

Using *linComb* the characterising equation for vectors reads:

$$v == linComb\ v\ e \tag{7.1}$$

Exercise 7.3. Using the elements defined above, sketch the isomorphism between an abstract vector space and its representation. Recall the definition of isomorphism in Section 4.2.3.

7.2 Linear transformations

As we have seen in earlier chapters, morphisms between structures are often important. Vector spaces are no different: if we have two vector spaces *Vector S G* and *Vector S G'* for the same set of scalars *S*, we can study functions $f : Vector\ S\ G \rightarrow Vector\ S\ G'$:

$$f\ v = f\ (v\ 0 \triangleleft e\ 0 + \ldots + v\ n \triangleleft e\ n)$$

It is particularly interesting to study functions which preserve the vector space structure: vector-space s. Such functions are more commonly called "linear maps", but to avoid unnecessary confusion with the Haskell *map* function we will refer to vector-space homomorphisms by the slightly less common name "linear transformation". Spelling out the homomorphism, the function *f* is a linear transformation if it maps the operations in *Vector S G* into operations in *Vector S G'* as follows:

$$f\ (u + v) = f\ u + f\ v$$
$$f\ (s \triangleleft u) = s \triangleleft f\ u$$

Because $v = linComb\ v\ e = (v\ 0 \triangleleft e\ 0 + \ldots + v\ n \triangleleft e\ n)$, we also have:

$$
\begin{aligned}
f\ v &= f\ (v\ 0 \triangleleft e\ 0 \quad + \ldots + v\ n \triangleleft e\ n) &&\{\text{- because } f \text{ is linear -}\} \\
&= \quad v\ 0 \triangleleft f\ (e\ 0) + \ldots + v\ n \triangleleft f\ (e\ n) &&\{\text{- by def. of } linComb \text{ -}\} \\
&= linComb\ v\ (f \circ e)
\end{aligned}
$$

But this means that we can determine the whole function $f : Vector\ S\ G \rightarrow Vector\ S\ G'$ on all vectors from just *f* on the base vectors: $f \circ e : G \rightarrow Vector\ S\ G'$, which has a much smaller domain. Let $m = f \circ e$. Then, for each *i*, the vector *m i* is the image of the canonical basis vector *e i* through *f*. Then

$$f\ v = linComb\ v\ m = v\ 0 \triangleleft m\ 0 + \ldots + v\ n \triangleleft m\ n$$

Each of the *m k* is a *Vector S G'*, as is the resulting *f v*. If we look at the component *g'* of *f v* we have

$$
\begin{aligned}
f\ v\ g' &= \{\text{- as above -}\} \\
(linComb\ v\ m)\ g' &= \{\text{- } linComb, (\triangleleft), (+) \text{ are all linear -}\} \\
linComb\ v\ (\lambda g \rightarrow m\ g\ g')
\end{aligned}
$$

That is, it suffices to know the behaviour of f on the basis vectors to know its behaviour on the whole vector space.

It is enlightening to compare the above sum with the standard vector-matrix multiplication. Let us define M as follows:

$$M = \begin{bmatrix} m\ 0 & \cdots & m\ n \end{bmatrix} \qquad \text{where } m : G \to Vector\ S\ G'$$

That is, the columns of M are $m\ 0$ to $m\ n$, or, in other words, the columns of M are $f\ (e\ i)$. Every $m\ k$ has *card* G' elements, and it has become standard to write $M\ i\ j$ to mean the ith element of the jth column, i.e., $M\ i\ j = m\ j\ i$, so that, if we denote the usual matrix-vector multiplication by *mulMV*:

$$(mulMV\ M\ v)\ i = linComb\ v\ (M\ i)$$

therefore, one has

$$
\begin{array}{lll}
(mulMV\ M\ v)\ i & = & \text{-- by def. of } mulMV \\
linComb\ v\ (M\ i) & = & \text{-- by def. of } M\ i\ j \\
linComb\ v\ (\lambda j \to m\ j\ i) & = & \text{-- earlier computation (linearity)} \\
f\ v\ i
\end{array}
$$

If we take *Matrix* to be just a synonym for functions of type $G \to Vector\ S\ G'$:

type *Matrix s g g′* $= g' \to Vector\ s\ g$

then we can implement matrix-vector multiplication as follows:

$mulMV :: (Finite\ g, Field\ s) \Rightarrow Matrix\ s\ g\ g' \to Vector\ s\ g \to Vector\ s\ g'$
$mulMV\ m\ (V\ v) = linComb\ v\ (transpose\ m)$

$transpose :: Matrix\ s\ i\ j \to Matrix\ s\ j\ i$
$transpose\ m\ i = V\ (\lambda j \to m\ j\ !\ i)$

In the terminology of the earlier chapters, we can see *Matrix s g g′* as a type of syntax and the linear transformation (of type $Vector\ S\ G \to Vector\ S\ G'$) as semantics. With this view, *mulMV* is just another evaluation function from syntax to semantics. However, again, given a fixed basis we have an isomorphism rather than a mere homomorphism: for a given linear transformation, the matrix representation is unique. Below we often write just an infix $(*)$ for *mulMV*.

Example Consider the multiplication of a matrix with a basis vector:

$$(M * e\ k)\ !\ i = (linComb\ (is\ k)\ (transpose\ M))\ !\ i = M\ i\ !\ k$$

i.e., *e k* extracts the *k*th column from *M* (hence the notation "e" for "extract").

We have seen how a linear transformation *f* can be fully described by a matrix of scalars, *M*. Similarly, in the opposite direction, given an arbitrary matrix *M*, we can define

$$f\, v = M * v$$

and obtain a linear transformation $f = (M*)$. Moreover $((M*) \circ e)\ g\ g' = M\ g'\ g$, i.e., the matrix constructed as above for *f* is precisely *M*.

In Exercise 7.9 you verify this by computing $((M*) \circ e)\ g\ g'$.

Therefore, every linear transformation is of the form $(M*)$ and every $(M*)$ is a linear transformation. There is a bijection between these two sets. Matrix-matrix multiplication is defined in order to ensure associativity (note here the overloading of the operator *):

$$(M' * M) * v = M' * (M * v)$$

that is, if we abstract over the vector *v* on both sides:

$$((M' * M)*) = (M'*) \circ (M*)$$

You may want to refer to Exercise 7.10, which asks you to work this out in detail, and Exercise 7.11 is about associativity of matrix-matrix multiplication.

A simple vector space is obtained for $G = ()$, the singleton index set. In this case, the vectors $s : () \rightarrow S$ are functions that can take exactly one value as argument, therefore they have exactly one value: *s* (), so they are isomorphic with *S*. But, for any $v : G \rightarrow S$, we have a function $fv : G \rightarrow (() \rightarrow S)$, namely

$$fv\ g\ () = v\ g$$

fv is similar to our *m* function above. The associated matrix is

$$M = \begin{bmatrix} m\ 0 & \cdots & m\ n \end{bmatrix} = \begin{bmatrix} fv\ 0 & \cdots & fv\ n \end{bmatrix}$$

having $n + 1$ columns (the dimension of *Vector G*) and one row (dimension of *Vector* ()). Let *w* :: *Vector S G*:

$$M * w = w\ 0 \triangleleft fv\ 0 + \ldots + w\ n \triangleleft fv\ n$$

$M * v$ and each of the *fv k* are "almost scalars": functions of type $() \rightarrow S$, thus, the only component of $M * w$ is

$$(M * w)\ () = w\ 0 * fv\ 0\ () + \ldots + w\ n * fv\ n\ ()$$
$$= w\ 0 * v\ 0 + \ldots + w\ n * v\ n$$

i.e., the scalar product of the vectors *v* and *w*.

Remark: We have not yet discussed the geometrical point of view.

7.3   Inner products

An important concept is the inner product between vectors. We define inner product space as a vector space equipped with an inner product, as follows:

> **class** *VectorSpace v s* $\Rightarrow$ *InnerSpace v s* **where**
> *inner* :: $v \to v \to s$

Inner products have (at least) two aspects. First, they yield a notion of how "big" a vector is, the *norm*.

> *sqNorm* :: *InnerSpace v s* $\Rightarrow$ $v \to s$
> *sqNorm v = inner v v*
>
> *norm* :: (*InnerSpace v a, Algebraic a*) $\Rightarrow$ $v \to a$
> *norm v* = $\sqrt{sqNorm\ v}$

Additionally, the inner product often serves as a measure of how much vectors are similar to (or correlated with) each other.

For two non-zero vectors *u* and *v*, we can define:

> *similarity u v = inner u v / norm u / norm v*

Dividing by the norms mean that *abs* (*similarity u v*) is at most 1 — the similarity is always in the interval $[-1, 1]$.

For example, in Euclidean spaces, one defines the inner product to be the product of the cosine of the angle between the vectors and their norms. Consequently, *similarity* is the cosine of the angle between vectors.

For this reason, one says that two vectors are orthogonal when their inner product is 0 — even in non-Euclidean spaces.

Dot product   An often used inner product is the dot product, defined as follows:[2]

> *dot* :: (*Field s, Finite g*) $\Rightarrow$ *Vector s g* $\to$ *Vector s g* $\to$ *s*
> *dot* (*V v*) (*V w*) = *linComb v w*

We should note that the dot product acts on the representations (syntax). This means that it will *change* depending on the basis chosen to represent vectors. Thus, the dot product is a syntactic concept, and it should be clearly identified

---

[2]This code is using the one-dimensional vector space instance defined in Chapter 7.

as such. This can be somewhat counterintuitive, because so far in this chapter it was fine to use representations (they were unique given the basis). To further confuse matters, in Euclidean spaces (which are often used as illustration) if the basis vectors are orthogonal, then the dot product coincides with the inner product. But, according to our methodology, one should start by defining a suitable inner product, and then check if the dot product is equivalent to it. See Section 7.4.3 for an example.

Orthogonal transformations An important subclass of the linear transformations are those which preserve the inner product.

inner $(f\ u)\ (f\ v) = inner\ u\ v$

In Euclidean spaces, such a transformation preserve angles. In the context of linear algebra they are either called orthogonal transformations (emphasising the preservation of angles) or unitary transformations (emphasising preservation of norms).[3]

Exercise 7.4. Can you express this condition as a homomorphism condition?

Such transformations necessarily preserve the dimension of the space (otherwise at least one basis vector would be squished to nothing and inner products involving it become zero). The corresponding matrices are square.

Exercise 7.5. Prove that orthogonal operators form a monoid with multiplication as an operator.

If angles are preserved what about distances? An isometry f is a distance-preserving transformation:

norm $(f\ v) = norm\ v$

We can prove that f is orthogonal iff. it is an isometry. The proof in the left-to-right direction is easy and left as an exercise. In the other direction one uses the equality:

$4 * inner\ u\ v = sqNorm\ (u+v) - sqNorm\ (u-v)$

In Euclidean spaces, this means that preserving angles and preserving distances go hand-in-hand.

---

[3]In today's mathematical vocabulary, the word "unitary" signals that a complex scalar field is used, whereas the word "orthogonal" signals that a real field is used, and that the space is Euclidean.

Orthogonal transformations enjoy many more useful properties — we have barely scratched the surface here. Among others, their rows (and columns) are orthogonal to each other. The are also invertible (and so they form a group), and the inverse is the given by (conjugate-) transpose of the matrix.

7.4   Examples of matrix algebra

7.4.1   Functions

A useful example of a vector space is the functions from $\mathbb{R}$ to $\mathbb{R}$. In terms of a *VectorSpace* instance we have:

> **instance** *VectorSpace* $(\mathbb{R} \to \mathbb{R})\ \mathbb{R}$ **where**
> $s \triangleleft f = (s*) \circ f$

Here $s \triangleleft f$ scales the function f by s pointwise.

Exercise 7.6. Verify the *VectorSpace* laws for the above instance.

An example of a linear transformation is the derivative. Indeed, we have already seen that $D\ (f + g) = D\ f + D\ g$. The equation $D\ (s \triangleleft f) = s \triangleleft D\ f$ is verified by expanding the definitions. Together, this means that the laws of linear transformations are verified.

7.4.2   Polynomials and their derivatives

In Chapter 5, we have represented polynomials of degree $n + 1$ by the list of their coefficients. This is the same representation as the vectors represented by $n + 1$ coordinates which we referred to in the introduction to this chapter. Indeed, polynomials of degree n form a vector space, and we could interpret that as $\{0, \ldots, n\} \to \mathbb{R}$ (or, more generally, *Field* $a \Rightarrow \{0, \ldots, n\} \to a$). The operations, $(+)$ for vector addition and $(\triangleleft)$ for vector scaling, are defined in the same way as they are for functions.

To give an intuition for the vector space it is useful to consider the interpretation of the canonical basis vectors. Recall that they are:

> $e\ i : \{0, \ldots, n\} \to \mathbb{R}; e\ i\ j = i\ 'is'\ j$

but how do we interpret them as polynomial functions?

When we represented a polynomial by its list of coefficients in Chapter 5, we saw that the polynomial function $\lambda x \to x\string^3$ could be represented as $[0, 0, 0, 1]$, where 1 is the coefficient of $x\string^3$.

This representation suggests to use as canonical basis vectors $e\ i$ the monomials $\lambda x \to x\string^i$. Representing the above list of coefficients as a vector is then a matter of converting lists to functions $\{0, \ldots, n\} \to \mathbb{R})$. This way, the vector $\lambda j \to$ **if** j == 3 **then** 1 **else** 0 is equal to $\lambda j \to 3$ '*is*' j or simply $e\ 3$. Any other polynomial function p equals the linear combination of monomials, and can therefore be represented as a linear combination of our basis vectors $e\ i$. For example, $p\ x = 2 + x\string^3$ is represented by $2 \triangleleft e\ 0 + e\ 3$.

The evaluator from the *Vector g s* representation to polynomial functions is as follows:

$$evalP :: Vector\ \mathbb{R}\ \{0, \ldots, n\} \to (\mathbb{R} \to \mathbb{R})$$
$$evalP\ (V\ v)\ x = sum\ (map\ (\lambda i \to v\ i * x\string^i)\ [0 .. n])$$

Let us now turn to the representation of the derivative of polynomials. We have already seen in the previous section that the *derive* function is a linear transformation. We also know that it takes polynomials of degree $n + 1$ to polynomials of degree n, and as such it is well defined as a linear transformation of polynomials too. Its representation can be obtained by applying the linear transformation to every basis vector:

$$M = [derive\ (e\ 0), derive\ (e\ 1), \ldots, derive\ (e\ n)]$$

where each *derive* $(e\ i)$ has length n. The vector $e\ (i + 1)$ represents $\lambda x \to x\string^(i + 1)$ and thus we want *derive* $(e\ (i + 1))$ to represent the derivative of $\lambda x \to x\string^(i + 1)$:

$$\begin{aligned}
&evalP\ (derive\ (e\ (i + 1))) && = \{\text{- by spec. -}\} \\
&D\ (evalP\ (e\ (i + 1))) && = \{\text{- by def. of } e, evalP \text{ -}\} \\
&D\ (\lambda x \to x\string^(i + 1)) && = \{\text{- derivative of a monomial -}\} \\
&\lambda x \to (i + 1) * x\string^i && = \{\text{- by def. of } e, evalP, (\triangleleft) \text{ -}\} \\
&evalP\ ((i + 1) \triangleleft (e\ i))
\end{aligned}$$

Thus

$$derive\ (e\ (i + 1)) = (i + 1) \triangleleft (e\ i)$$

Also, the derivative of $evalP\ (e\ 0) = \lambda x \to 1$ is $\lambda x \to 0$ and thus *derive* $(e\ 0)$ is the zero vector:

$$derive\ (e\ 0) = 0$$

Example: $n + 1 = 3$:

$$M = \begin{bmatrix} 0 & 1 & 0 \\ 0 & 0 & 2 \end{bmatrix}$$

Take the polynomial function $p\ x = 1 + 2 * x + 3 * x\hat{}2$ as a vector

$$v = \begin{bmatrix} 1 \\ 2 \\ 3 \end{bmatrix}$$

and we have

$$M * v = \left[\begin{bmatrix} 0 \\ 0 \end{bmatrix} \begin{bmatrix} 1 \\ 0 \end{bmatrix} \begin{bmatrix} 0 \\ 2 \end{bmatrix} \right] * \begin{bmatrix} 1 \\ 2 \\ 3 \end{bmatrix} = \begin{bmatrix} 2 \\ 6 \end{bmatrix}$$

representing the polynomial function $p'\ x = 2 + 6 * x$.

As an interesting follow-up, Exercise 7.12 asks you to write the (infinite-dimensional) matrix representing D for power series. Similarly, in Exercise 7.13 you compute the matrix In associated with integration of polynomials.

7.4.3 *Inner product for functions and Fourier series

We said before that the inner product yields a notion of norm and similarity. Can we use the dot product as inner product for power series (if the basis is $e\ i = x\hat{}i$)? We could, but then it would not be very useful. For example, it would not yield a useful notion of similarity between the represented function. To find a more useful inner product, we can return to the semantics of power series in terms of functions. But for now we consider them over the restricted domain $I = [-\pi, \pi]$.

Assume for a moment that we would define the inner product of functions u and v as follows:

$innerF\ u\ v = \int_I (eval\ u\ x) * (eval\ v\ x)\ dx$

Then, the norm of a function would be a measure of how far it gets from zero, using a quadratic mean. Likewise, the corresponding similarity measure corresponds to how much the functions "agree" on the interval. That is, if the signs of $eval\ u$ and $eval\ v$ are the same on a sub-interval I then the integral is positive on I, and negative if they are different.

As we suspected, using $inner = innerF$, the straightforward representation of polynomials as list of coefficients is not an orthogonal basis. There is, for example, a positive correlation between the canonical vectors x and $x\hat{}3$.

If we were instead using a set of basis polynomials b_n which are orthogonal using the above definition of *inner*, then we could simply let *inner = dot*, and this would be a lot more efficient than to compute the integral by the following series of steps: 1) compute the product using *mulPoly*, 2) integrate using *integ*, 3) use *eval* on the end points of the domain.

Let us consider as a basis the functions $b_n\ x = sin\ (n * x)$, and prove that they are orthogonal.

We first use trigonometry to rewrite the product of basis vectors (call it $f_{ij}\ x$):

$$f_{ij}\ x$$
$$= 2 * (b_i\ x * b_j\ x)$$
$$= 2 * sin\ (i * x) * sin\ (j * x)$$
$$= cos\ ((i - j) * x) - cos\ ((i + j) * x)$$

Assuming $i \neq j$, we can take the indefinite integral, and find (call it $F_{ij}\ x$):

$$F_{ij}\ x = sin\ ((i - j) * x)\ /\ (i - j) - sin\ ((i + j) * x)\ /\ (i + j) + K$$

Note that we only need the value of this function at the interval end-points: $-\pi$ and π. But $sin\ (k * \pi) = 0$ for any integer k, which means that $F_{ij}\ (-\pi) = F_{ij}\ \pi = K$. Taking the definite integral over the domain I yields:

$$2 * (inner\ b_i\ b_j) = F_{ij}\ \pi - F_{ij}\ (-\pi) = K - K = 0$$

and thus $inner\ b_i\ b_j = 0$

We can now compute *sqNorm* of b_i. Trigonometry says:

$$2 * (b_i\ x * b_i\ x)$$
$$= 2 * sin\ (i * x) * sin\ (i * x)$$
$$= cos\ (0 * x) - cos\ (2 * i * x)$$
$$= 1 - cos\ (2 * i * x)$$

When taking the integral on I, the cosine disappears using the same argument as before, and there remains: $2 * sqNorm\ b_i = 2\ \pi$. Thus to normalise the basis vectors we need to scale them by $1\ /\ \sqrt{\pi}$. In sum $b_i\ x = sin\ (i * x)\ /\ \sqrt{\pi}$ is an orthonormal basis:

$$b_i\ 'innerF'\ b_j = is\ i\ j$$

As interesting as it is, this basis does not cover all functions over I. To start, *eval* $b_i\ 0 == 0$ for every i, and thus linear combinations can only ever be zero at the origin.

But if we were to include $cos \ (n * x) \ / \ \sqrt{\pi}$ in the set of basis vectors, it would remain orthogonal, and the space would cover all periodic functions with period 2π. A representation of a function in this basis is called the Fourier series. Let us define a meaningful index (G) for the basis:

> **data** *Periodic* **where**
>   *Sin* :: $\mathbb{N}_{>0}$ $\rightarrow$ *Periodic*
>   *Cos* :: $\mathbb{N}$ $\rightarrow$ *Periodic*
>   **deriving** *Eq*

For example, the function $f \ x = 3 * sin \ x + cos \ (2 * x) - 1$ is represented by the vector $v = 3 \triangleleft e \ (Sin \ 1) + e \ (Cos \ 2) - e \ (Cos \ 0)$ which can also be written:

> v :: *Vector* $\mathbb{R}$ *Periodic*
> $v = V \ vf$
>   **where** $vf \ (Sin \ 1) = 3$
> $vf \ (Cos \ 2) = 1$
> $vf \ (Cos \ 0) = -1$
> $vf \ \_ \qquad = 0$

A useful property of an orthonormal basis is that its representation as coefficients can be obtained by taking the inner product with each basis vectors. Indeed, using Eq. (7.1) as a starting point, we can calculate:[4]

> $\quad v \qquad\qquad$ == *linComb* $v \ b$
> $\Rightarrow v \qquad\qquad$ == *sum* $[v_i \triangleleft b_i \mid i \leftarrow finiteDomain]$
> $\Rightarrow v$ '*inner*' b_j == *sum* $[v_i \triangleleft b_i \mid i \leftarrow finiteDomain]$ '*inner*' b_j
> $\Rightarrow v$ '*inner*' b_j == *sum* $[v_i \triangleleft (b_i \ 'inner' \ b_j) \mid i \leftarrow finiteDomain]$
> $\Rightarrow v$ '*inner*' b_j == *sum* $[v_i \triangleleft is \ i \ j \mid i \leftarrow finiteDomain]$
> $\Rightarrow v$ '*inner*' b_j == v_j

Thus, in our application, given a periodic function f, one can compute its Fourier series by taking the *innerF* product of it with each of $sin \ (n * x) \ / \ \sqrt{\pi}$ and $cos \ (n * x) \ / \ \sqrt{\pi}$.

Exercise 7.7. Derive *derive* for this representation.

7.4.4   Simple deterministic systems (transition systems)

Simple deterministic systems are given by endofunctions on a finite set *next* : $G \rightarrow G$. They can often be conveniently represented as a graph, for example

---

[4]The proof can be easily adapted to infinite sums.

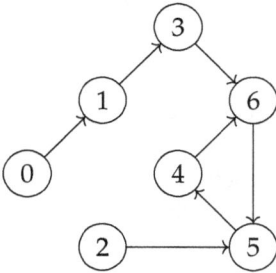

Here, $G = \{0,\dots,6\}$. A node in the graph represents a state. A transition $i \to j$ means *next* $i = j$. Since *next* is an endofunction, every node must be the source of exactly one arrow.

We can take as vectors the characteristic functions of subsets of G, i.e., $G \to \{0,1\}$. Now, $\{0,1\}$ is not a field with respect to the standard arithmetical operations (it is not even closed with respect to addition), and the standard trick to workaround this issue is to extend the type of the functions to $\mathbb{R}$.

The canonical basis vectors are, as usual, $e\,i = V\ (is\ i)$. Each $e\,i$ is the characteristic function of a singleton set, $\{i\}$.

We can interpret $e\ (next\ 0),\dots,e\ (next\ 6)$ as the images of the basis vectors $e\,0,\dots,e\,6$ of *Vector* $\mathbb{R}\ G$ under the transformation

$$f :: Vector\ \mathbb{R}\ G \to Vector\ \mathbb{R}\ G$$
$$f\ (e\,i) = e\ (next\ i)$$

To write the matrix associated to f, we have to compute what vector is associated to each canonical basis vector:

$$M = \begin{bmatrix} f\ (e\,0) & \cdots & f\ (e\,n) \end{bmatrix}$$

Therefore:

$$
M = \begin{array}{c}
 \\ r_0 \\ r_1 \\ r_2 \\ r_3 \\ r_4 \\ r_5 \\ r_6
\end{array}
\begin{array}{ccccccc}
c_0 & c_1 & c_2 & c_3 & c_4 & c_5 & c_6 \\
\left(\begin{array}{ccccccc}
0 & 0 & 0 & 0 & 0 & 0 & 0 \\
1 & 0 & 0 & 0 & 0 & 0 & 0 \\
0 & 0 & 0 & 0 & 0 & 0 & 0 \\
0 & 1 & 0 & 0 & 0 & 0 & 0 \\
0 & 0 & 0 & 0 & 0 & 1 & 0 \\
0 & 0 & 1 & 0 & 0 & 0 & 1 \\
0 & 0 & 0 & 1 & 1 & 0 & 0
\end{array}\right)
\end{array}
$$

Notice that row 0 and row 2 contain only zero, as one would expect from the graph of *next*: no matter where we start from, the system will never reach node 0 or node 2.

Starting with a canonical basis vector $e\ i$, we obtain $M*e\ i = f\ (e\ i)$, as we would expect. The more interesting thing is if we start with something different from a basis vector, say $[0,0,1,0,1,0,0]$ == $e\ 2 + e\ 4$. We obtain $\{f\ 2, f\ 4\} = \{5,6\}$, the image of $\{2,4\}$ through f. In a sense, we can say that the two transitions happened in parallel. But that is not quite accurate: if we start with $\{3,4\}$, we no longer get the characteristic function of $\{f\ 3, f\ 4\} = \{6\}$, instead, we get a vector that does not represent a characteristic function at all: $[0,0,0,0,0,0,2] = 2 \triangleleft e\ 6$.

In general, if we start with an arbitrary vector, we can interpret this as starting with various quantities of some unspecified material in each state, simultaneously. If f were injective, the respective quantities would just get shifted around, but in our case, we get a more general behaviour.

What if we do want to obtain the characteristic function of the image of a subset? In that case, we need to use other operations than the standard arithmetical ones, for example *min* and *max*.

However, $(\{0,1\}, max, min)$ is not a field, and neither is $(\mathbb{R}, max, min)$. This means that we do not have a vector space, but rather a *module* (a generalisation of vector space). One can still do a lot with modules: for example the definition of matrix multiplication only demands a *Ring* rather than a *Field* (and none of the *VectorSpace* laws demand scalar division). Therefore, having just a module is not a problem if all we want is to compute the evolutions of possible states, but we cannot apply most of the deeper results of linear algebra.[5]

In the example above, we have:

```
newtype G = G Int deriving (Eq, Show)
instance Bounded G where minBound = G 0; maxBound = G 6
instance Enum G    where toEnum = G; fromEnum (G g) = g
```

Note that the *Ring G* instance is given just for convenient notation (integer literals): vector spaces in general do not rely on any numeric structure on the indices (G). The transition function has type $G \to G$ and the following implementation:

```
next₁ :: G → G
next₁ (G 0) = G 1;    next₁ (G 1) = G 3;
next₁ (G 2) = G 5;    next₁ (G 3) = G 6;
next₁ (G 4) = G 6;    next₁ (G 5) = G 4;
next₁ (G 6) = G 5
```

---

[5]For instance, such a deeper result would give ways to easily compute the stable states of a dynamic system.

Its associated matrix is

$m\ g'$	{- m is the matrix version of f -}
$V\ (\lambda g \to (f\ (e\ g)))$ $!\,g')$	{- by the spec. of f -}
$V\ (\lambda g \to (e\ (next_1\ g)))$ $!\,g')$	{- by def. of e -}
$V\ (\lambda g \to (V\ (is\ (next_1\ g))))\,!\,g')$	{- by def. of (!) -}
$V\ (\lambda g \to is\ (next_1\ g)\ g')$	{- is is symmetric -}
$V\ (\lambda g \to is\ g'\ (next_1\ g))$	{- by def. of ($\circ$) -}
$V\ (is\ g' \circ next_1)$	

Thus we can implement m as:

$$m_1 :: Ring\ s \Rightarrow G \to Vector\ s\ G$$
$$m_1\ g' = V\ (is\ g' \circ next_1)$$

7.4.5 Non-deterministic systems

Another interpretation of the application of M to characteristic functions of a subset is the following: assuming that all I know is that the system is in one of the states of the subset, where can it end up after one step? (This assumes the *max-min* algebra as above.)

The general idea for non-deterministic systems, is that the result of applying the step function a number of times from a given starting state is a list of the possible states one could end up in.

In this case, the uncertainty is entirely caused by the fact that we do not know the exact initial state. However, there are cases in which the output of f is not known, even when the input is known. Such situations are modelled by endo-relations: $R : G \to G$, with $g\ R\ g'$ if g' is a potential successor of g. Endo-relations can also be pictured as graphs, but the restriction that every node should be the source of exactly one arrow is lifted. Every node can be the source of zero, one, or many arrows.

For example:

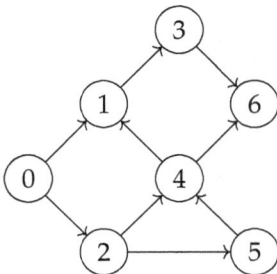

Now, starting in 0 we might end up either in 1 or 2 (but not both!). Starting in 6, the system breaks down: there is no successor state.

The matrix associated to R is built in the same fashion: we need to determine what vectors the canonical basis vectors are associated with:

$$
M = \begin{array}{c} \\ r_0 \\ r_1 \\ r_2 \\ r_3 \\ r_4 \\ r_5 \\ r_6 \end{array}
\begin{array}{c} \begin{array}{ccccccc} c_0 & c_1 & c_2 & c_3 & c_4 & c_5 & c_6 \end{array} \\
\left(\begin{array}{ccccccc}
0 & 0 & 0 & 0 & 0 & 0 & 0 \\
1 & 0 & 0 & 0 & 1 & 0 & 0 \\
1 & 0 & 0 & 0 & 0 & 0 & 0 \\
0 & 1 & 0 & 0 & 0 & 0 & 0 \\
0 & 0 & 1 & 0 & 0 & 1 & 0 \\
0 & 0 & 1 & 0 & 0 & 0 & 0 \\
0 & 0 & 0 & 1 & 1 & 0 & 0
\end{array}\right)
\end{array}
$$

In Exercise 7.14 you are asked to start with $e\,2 + e\,3$ and iterate a number of times, to get a feeling for the possible evolutions.

Implementation   The transition relation is given by:

$$
\begin{aligned}
&f_2 :: G \rightarrow (G \rightarrow Bool) \\
&f_2\ (G\ 0)\ (G\ g) = g \mathrel{==} 1 \vee g \mathrel{==} 2 \\
&f_2\ (G\ 1)\ (G\ g) = g \mathrel{==} 3 \\
&f_2\ (G\ 2)\ (G\ g) = g \mathrel{==} 4 \vee g \mathrel{==} 5 \\
&f_2\ (G\ 3)\ (G\ g) = g \mathrel{==} 6 \\
&f_2\ (G\ 4)\ (G\ g) = g \mathrel{==} 1 \vee g \mathrel{==} 6 \\
&f_2\ (G\ 5)\ (G\ g) = g \mathrel{==} 4 \\
&f_2\ (G\ 6)\ (G\ g) = False
\end{aligned}
$$

It has the associated matrix:

$$
m_2\ g' = V\ (\lambda g \rightarrow f_2\ g\ g')
$$

Even though *Bool* is not a *Field* (not even a *Ring*) the computations we need go through with these instances:

> **instance** *Additive*   *Bool* **where** $zero = False; (+) = (\vee)$
> **instance** *Multiplicative Bool* **where** $one = True; (*) = (\wedge)$
>
> **instance** *AddGroup Bool* **where** $negate = error$ `"negate: not used"`
> **instance** *MulGroup Bool* **where** $recip = id$

As a test we compute the state after one step from "either 3 or 4":

$$
\begin{aligned}
&t2' = mulMV\ m_2\ (e\ (G\ 3) + e\ (G\ 4)) \\
&t_2 = toL\ t2' \quad \text{-- } [False, True, False, False, False, False, True]
\end{aligned}
$$

7.4.6 Stochastic systems

Quite often, we have more information about the transition to possible future states. In particular, we can have *probabilities* of these transitions. For example

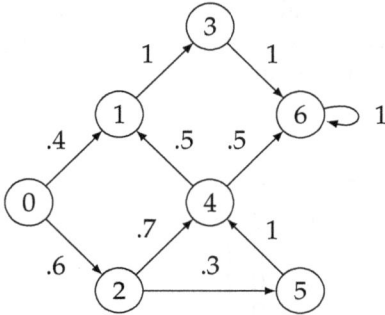

One could say that this case is a generalisation of the previous one, in which we can take all probabilities to be equally distributed among the various possibilities. While this is plausible, it is not entirely correct. For example, we have to introduce a transition from state 6 above. The nodes must be sources of *at least* one arrow.

In the case of the non-deterministic example, the "legitimate" inputs were characteristic functions, i.e., the "vector space" was $G \to \{0,1\}$ (the quotes are necessary because, as discussed, the target is not a field). In the case of stochastic systems, the inputs will be *probability distributions* over G, that is, functions $p : G \to [0,1]$ with the property that

$$sum\ [p\ g \mid g \leftarrow G] = 1$$

If we know the current probability distributions over states, then we can compute the next one by using the *total probability formula*, which can be expressed as

$$p\ a = sum\ [p\ (a \mid b) * p\ b \mid b \leftarrow G]$$

We study probability extensively in Chapter 9, but for now, let's just remark that the notation is extremely suspicious. The "argument" $(a \mid b)$, which is usually read "a, given b", is clearly not of the same type as a or b, so cannot really be an argument to p. Additionally, the $p\ a$ we are computing with this formula is not the $p\ a$ which must eventually appear in the products on the right hand side.

In any case, at this stage, what we need to know is that the conditional probability $p\ (a \mid b)$ gives us the probability that the next state is a, given that the current state is b. But this is exactly the information summarised in the graphical representation. Moreover, it can be shown that the total probability formula is identical to a matrix-vector multiplication.

As usual, we write the associated matrix by looking at how the canonical basis vectors are transformed. In this case, the canonical basis vector $e\ i = \lambda j \rightarrow$ i 'is' j is the probability distribution *concentrated* in i. This means that the probability to be in state i is 100% and the probability of being anywhere else is 0.

$$
M = \begin{array}{c}
 \\
r_0 \\
r_1 \\
r_2 \\
r_3 \\
r_4 \\
r_5 \\
r_6
\end{array}
\begin{pmatrix}
c_0 & c_1 & c_2 & c_3 & c_4 & c_5 & c_6 \\
0 & 0 & 0 & 0 & 0 & 0 & 0 \\
.4 & 0 & 0 & 0 & .5 & 0 & 0 \\
.6 & 0 & 0 & 0 & 0 & 0 & 0 \\
0 & 1 & 0 & 0 & 0 & 0 & 0 \\
0 & 0 & .7 & 0 & 0 & 1 & 0 \\
0 & 0 & .3 & 0 & 0 & 0 & 0 \\
0 & 0 & 0 & 1 & .5 & 0 & 1
\end{pmatrix}
$$

As before, a good exercise is to explore the evolution of the system. For example, in Exercise 7.15 you are asked how many steps you need to take before the probability is concentrated in state 6 starting from state 0? And in Exercise 7.16 you are tasked with implementing the example by defining the transition function (giving the probability of getting to g' from g)

$$f_3 :: G \rightarrow Vector\ \mathbb{R}\ G$$

and the associated matrix

$$m_3 :: G \rightarrow Vector\ \mathbb{R}\ G$$

7.4.7 *Quantum Systems

Instead of real numbers for probabilities, we could consider using complex numbers — one then speaks of "amplitudes". An amplitude represented by a complex number z is converted to a probability by taking the square of the modulus of z:

$$p\ z = conj\ z * z$$

We can then rewrite the law of total probability as follows:

$$
\begin{aligned}
&sum\ [\,p\,!\,i \mid i \leftarrow finiteDomain\,] \\
={}&sum\ [\,conj\ (z\,!\,i) * (z\,!\,i) \mid i \leftarrow finiteDomain\,] \\
={}&inner\ z\ z
\end{aligned}
$$

Indeed, for spaces with complex scalars, one should conjugate coefficients (of an orthonormal basis) when computing the inner products. Hence, rather conveniently, the law of total probability is replaced by conservation of the norm of state vectors. In particular, norms are conserved if the transition matrix is unitary.

The unitary character of the transition matrix defines valid systems from the point of view of quantum mechanics. Because all unitary matrices are invertible, it follows that all quantum mechanical systems have an invertible dynamics. Furthermore, the inverted matrix is also unitary, and therefore the inverted system is also valid as a quantum dynamical system.

Here is an example unitary matrix

$$
M = \begin{array}{c} \\ r_0 \\ r_1 \\ r_2 \\ r_3 \\ r_4 \\ r_5 \\ r_6 \end{array}
\begin{pmatrix}
\overset{c_0}{0} & \overset{c_1}{0} & \overset{c_2}{1} & \overset{c_3}{0} & \overset{c_4}{0} & \overset{c_5}{0} & \overset{c_6}{0} \\
1 & 0 & 0 & 0 & 0 & 0 & 0 \\
0 & 1 & 0 & 0 & 0 & 0 & 0 \\
0 & 0 & 0 & \sqrt{2}/2 & -\sqrt{2}/2 & 0 & 0 \\
0 & 0 & 0 & \sqrt{2}/2 & \sqrt{2}/2 & 0 & 0 \\
0 & 0 & 0 & 0 & 0 & 1/2 & \sqrt{3}/2 \\
0 & 0 & 0 & 0 & 0 & \sqrt{3}/2 & -1/2
\end{pmatrix}
$$

In this example the amplitudes of states 0, 1, and 2 are permuted at every step. States 3 and 4 get mixed into one another, and one can note that the sign of their amplitudes may get inverted. A similar situation happens between states 5 and 6, but at a higher rate.

7.5 *Monadic dynamical systems

This section is meant to give perspective for the readers who are already familiar with monads. Even though it can be safely skipped, it presents a useful unified view of the previous sections which could help understanding the material.

All the examples of dynamical systems we have seen in the previous section have a similar structure. They work by taking a state (which is one of the generators) and return a structure of possible future states of type G:

- deterministic: there is exactly one possible future state: we take an element of G and return an element of G. The transition function has the type $f : G \rightarrow G$, the structure of the target is just G itself.

- non-deterministic: there is a set of possible future states, which we have implemented as a characteristic function $G \to \{0,1\}$. The transition function has the type $f : G \to (G \to \{0,1\})$. The structure of the target is the *powerset* of G.

- stochastic: given a state, we compute a probability distribution over possible future states. The transition function has the type $f : G \to (G \to \mathbb{R})$, the structure of the target is the probability distributions over G.

- quantum: given an observable state, we compute a superposition of possible orthogonal future states.

Therefore:

- deterministic: $f : G \to Id\ G$

- non-deterministic: $f : G \to Powerset\ G$, where $Powerset\ G = G \to \{0,1\}$

- stochastic: $f : G \to Prob\ G$, where $Prob\ G = G \to [0,1]$

- quantum: $f : G \to Super\ G$, where $Super\ G = G \to Complex$. (Additionally f must be invertible)

We have represented the elements of the various structures as vectors. We also had a way of representing, as structures of possible states, those states that were known precisely: these were the canonical basis vectors $e\ i$. Due to the nature of matrix-vector multiplication, what we have done was in effect:

$$M * v \quad \text{-- } v \text{ represents the current possible states}$$
$$= \{\text{- } v \text{ is a linear combination of the basis vectors -}\}$$
$$M * (v\ 0 \vartriangleleft e\ 0 + \ldots + v\ n \vartriangleleft e\ n)$$
$$= \{\text{- homomorphism -}\}$$
$$v\ 0 \vartriangleleft (M * e\ 0) + \ldots + v\ n \vartriangleleft (M * e\ n)$$
$$= \{\text{- } e\ i \text{ represents the known current state } i, \text{ therefore } M * e\ i = f\ i \text{ -}\}$$
$$v\ 0 \vartriangleleft f\ 0 + \ldots + v\ n \vartriangleleft f\ n$$

So, we apply f to every state, as if we were starting from precisely that state, obtaining the possible future states starting from that state, and then collect all these hypothetical possible future states in some way that takes into account the initial uncertainty (represented by $v\ 0$, ..., $v\ n$) and the nature of the uncertainty (the specific $(+)$ and $(\vartriangleleft)$).

If you examine the types of the operations involved

$e : G \to Possible\ G$

and

$flip\ (*) : Possible\ G \to (G \to Possible\ G) \to Possible\ G$

you see that they are very similar to the monadic operations

$return : g \to m\ g$
$(\ggg) : m\ g \to (g \to m\ g') \to m\ g'$

which suggests that the structure of possible future states might be monadic. Indeed, that is the case.

Since we implemented all these as matrix-vector multiplications, this raises the question: is there a monad underlying matrix-vector multiplication, such that the above are instances of it (obtained by specialising the scalar type S)? The answer is yes, up to a point, as we shall see in the next section.

Exercise 7.8 (*Hard). Write *Monad* instances for *Id*, *Powerset*, *Prob*, *Super*.

7.5.1 *The monad of linear algebra

Haskell *Monad*s, just like *Functor*s, require *return* and $\ggg$ to be defined for every type. This will not work, in general. Our definition will work for *finite types* only.

```
class FinFunc f where
    func :: (Finite a, Finite b) ⇒ (a → b) → f a → f b
class FinMon f where
    embed :: Finite a ⇒ a → f a
    bind   :: (Finite a, Finite b) ⇒ f a → (a → f b) → f b
```

The idea is that vectors on finite types are finite functors and monads:

```
instance Ring s ⇒ FinFunc (Vector s) where
    func f (V v) = V (λg' → sum [v g | g ← finiteDomain, g' == f g])
instance Field s ⇒ FinMon (Vector s) where
    embed = embedFinM
    bind   = bindFinM
embedFinM :: (Eq a, Ring s) ⇒ a → Vector s a
embedFinM g = V (is g)
bindFinM :: (Field s, Finite a) ⇒ Vector s a → (a → Vector s b) → Vector s b
bindFinM (V v) f = V (λg' → linComb v (λg → f g ! g'))
```

Note that, if $v :: Vector\ S\ G$ and $f :: G \rightarrow Vector\ S\ G'$ then both *func f v* and *bind v f* are of type *Vector S G'*. How do these operations relate to linear algebra and matrix-vector multiplication?

Remember that *e g* is that vector whose components are zero except for the *g*th one which is one. In other words

$$e\ g = V\ (is\ g) = embed\ g$$

and thus *embed = e*. In order to understand how matrix-vector multiplication relates to the monadic operations, remember that matrices are just functions of type $G \rightarrow Vector\ S\ G'$:

type *Matrix s g g' = g' → Vector s g*

According to our earlier definition, we can rewrite matrix-vector multiplication in terms of *linComb*

> $mulMV\ m\ (V\ v)$
> $= \{\text{- earlier definition -}\}$
> $linComb\ v\ (transpose\ m)$

Now we have:

> $mulMV\ (transpose\ m)\ (V\ v)$
> $= \{\text{- def. of } mulMV \text{ -}\}$
> $linComb\ v\ (transpose\ (transpose\ m))$
> $= \{\text{- Property of } transpose \text{ -}\}$
> $linComb\ v\ m$
> $= \{\text{- } linComb\text{-}V \text{ lemma -}\}$
> $V\ (\lambda i \rightarrow linComb\ v\ (\lambda j \rightarrow m\ j\ !\ i))$
> $= \{\text{- def. of } bind \text{ -}\}$
> $bind\ (V\ v)\ m$

Thus we see that *bind v f* is "just" a matrix-vector multiplication.

The *linComb-V* lemma says that for $a :: j \rightarrow s$ and $v :: j \rightarrow Vector\ s\ i$ we have $linComb\ a\ v\ ==\ V\ (\lambda i \rightarrow linComb\ a\ (\lambda j \rightarrow v\ j\ !\ i))$. The proof uses the definitions of *linComb*, ($\triangleleft$), and the *Additive* instances for *Vector* and functions but is omitted here for brevity.

7.6   Associated code

Conversions and *Show* functions so that we can actually see our vectors.

toL :: *Finite g* $\Rightarrow$ *Vector s g* $\rightarrow$ [*s*]
toL (*V v*) = *map v finiteDomain*

instance (*Finite g, Show s*) $\Rightarrow$ *Show* (*g* $\rightarrow$ *s*)   **where** *show* = *showFun*
instance (*Finite g, Show s*) $\Rightarrow$ *Show* (*Vector s g*) **where** *show* = *showVec*

showVec :: (*Finite g, Show s*) $\Rightarrow$ *Vector s g* $\rightarrow$ *String*
showVec (*V v*) = *showFun v*

showFun :: (*Finite a, Show b*) $\Rightarrow$ (*a* $\rightarrow$ *b*) $\rightarrow$ *String*
showFun f = *show* (*map f finiteDomain*)

7.7   Exercises

Some exercises were inlined in the chapter text, and here are a few more.

Exercise 7.9. Compute $((M*) \circ e)\, g\, g'$.

Exercise 7.10. Matrix-matrix multiplication is defined in order to ensure a homomorphism from $(*)$ to $(\circ)$.

$$\forall M.\ \forall M'.\ ((M' * M)*) \mathrel{==} (M'*) \circ (M*)$$

or in other words

$$H_2((*), (*), (\circ))$$

Work out the types and expand the definitions to verify that this claim holds. Note that one $(*)$ is matrix-vector multiplication and the other is matrix-matrix multiplication.

Exercise 7.11. Show that matrix-matrix multiplication is associative.

Exercise 7.12. With $G = \mathbb{N}$ for the set of indices, write the infinite-dimensional matrix representing D for power series.

Exercise 7.13. Write the matrix I_n associated with integration of polynomials of degree n.

Exercise 7.14. In the context of Section 7.4.5: start with $v_0 = e\, 2 + e\, 3$ and iterate $M*$ a number of times, to get a feeling for the possible evolutions. What do you notice? What is the largest number of steps you can make before the result is the origin vector (just zero)?

Now change M to M' by inverting the arrow from 2 to 4 and repeat the exercise. What changes? Can you prove it?

Exercise 7.15. In the context of the example matrix M in Section 7.4.6: starting from state 0, how many steps do you need to take before the probability is concentrated in state 6?

Reverse again the arrow from 2 to 4 (so that $2 \to 5$ has probability 1, $4 \to 2$ probability 0.7, $4 \to 6$ and $4 \to 1$ have probability 0.15 each). What can you say about the long-term behaviour of the system now?

Exercise 7.16. In the context of the example matrix M in Section 7.4.6: implement the example. You will need to define the transition function of type $G \to (G \to [0,1])$ returning the probability of getting from g to g', and the associated matrix.

Exercise 7.17. *From exam 2017-03-14*

Consider a non-deterministic system with a transition function $f : G \to [G]$ (for $G = \{0 . . 5\}$) represented in the following graph

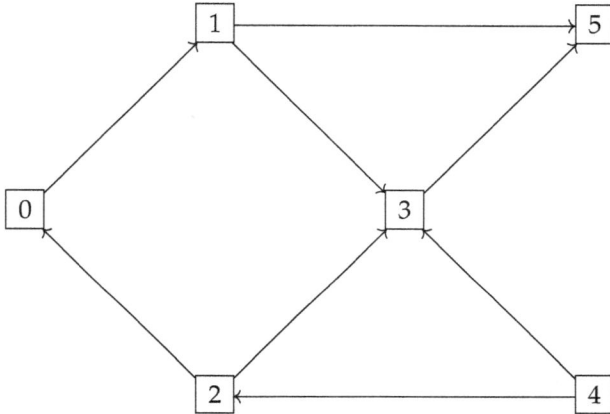

The transition matrix can be given the type $m :: G \to (G \to Bool)$ and the canonical vectors have type $e\ i :: G \to Bool$ for i in G.

1. (General questions.) What do the canonical vectors represent? What about non-canonical ones? What are the operations on *Bool* used in the matrix-vector multiplication?

2. (Specific questions.) Write the transition matrix m of the system. Compute, using matrix-vector multiplication, the result of three steps of the system starting in state 2.

Chapter 8

Exponentials and Laplace

8.1 The Exponential Function

In one of the classical analysis textbooks, Rudin [1987] starts with a prologue on the exponential function. The first sentence is

> This is undoubtedly the most important function in mathematics.

Rudin goes on

> It is defined, for every complex number z, by the formula
>
> $$exp(z) = \sum_{n=0}^{\infty} \frac{z^n}{n!}$$

We, on the other hand, have defined the exponential function as the solution of a differential equation, which can be represented by a power series:

expx :: *Field a* $\Rightarrow$ *PowerSeries a*
expx = *integ* 1 *expx*

and approximated by

expf :: *Field a* $\Rightarrow$ *a* $\rightarrow$ *a*
expf = *evalPS* 100 *expx*

It is easy to see, using the definition of *integ* that the power series *expx* is, indeed

$$expx = Poly\ [1, 1\ /\ 1, 1\ /\ 2, 1\ /\ 6, \ldots, 1\ /\ (1*2*3*\ldots*n), ..]$$

We can compute the exponential for complex values if we can give an instance of *Field* for complex numbers. We use the datatype *Data.Complex* from the Haskell standard library, which is isomorphic to the implementation from Chapter 1. In *Data.Complex* a complex value z is represented by two values, the real and the imaginary part, connected by an infix constructor: $z = re \mathbin{:\!\!+} im$.

i :: *Ring a* $\Rightarrow$ *Complex a*
i = *zero* :+ *one*

Therefore we can define, for example, the exponential of the imaginary unit:

ex1 :: *Field a* $\Rightarrow$ *Complex a*
ex1 = *expf i*

And we have *ex1* == 0.5403023058681398 :+ 0.8414709848078965. Observe at the same time:

cosf 1 = 0.5403023058681398
sinf 1 = 0.8414709848078965

and therefore *expf i* == *cosf* 1 :+ *sinf* 1. This is no coincidence, as we shall see.

First we define a helper function *compScale* which scales the input to a function by a factor. It is specified by *eval* (*compScale c as*) x = *eval as* ($c * x$). Then we define the power series of $f(x) = e^{ix}$ as *compScale i expx*.

```
compScale :: Ring a ⇒ a → PowerSeries a → PowerSeries a
compScale c (Poly as) = Poly (zipWith (*) as (iterate (c*) 1))
type PSC a = PowerSeries (Complex a)
expix :: Field a ⇒ PSC a
expix = compScale i expx
cosxisinx :: Field a ⇒ PSC a
cosxisinx = cosx + Poly [i] * sinx
```

As the code is polymorphic in the underlying number type, we can use rationals to be able to test for equality without rounding problems. We can see that every second coefficient is real and every second is imaginary:

ex2, *ex2'* :: *Field a* $\Rightarrow$ *PSC a*
ex2 = *takePoly* 8 *expix*

ex2' = *takePoly* 8 *cosxisinx*
test2 :: *Bool*
test2 = *ex2* == (*ex2'* :: *PSC Rational*)
check2 :: *Bool*
check2 = *ex2* == *coeff2*

coeff2 :: *PSC Rational*
coeff2 = *Poly* [1   :+ 0,   0 :+   1,
          (− 1 / 2)   :+ 0,   0 :+ (−1 / 6),
           1 / 24   :+ 0,   0 :+   1 / 120,
          (− 1 / 720) :+ 0,   0 :+ (−1 / 5040)]

We can see that the real part of this series is the same as

ex2R :: *Poly Rational*
ex2R = *takePoly* 8 *cosx*

and the imaginary part is the same as

ex2I :: *Poly Rational*
ex2I = *takePoly* 8 *sinx*

Therefore, the coefficients of *cosx* are

$$[1, 0, -1 / 2!, 0, 1 / 4!, 0, -1 / 6!, \ldots]$$

In other words, the power series representation of the coefficients for *cos* is

cosa (2 ∗ *n*)   = (−1)^*n* / (2 ∗ *n*) !
cosa (2 ∗ *n* + 1) = 0

and the terms of *sinx* are

$$[0, 1, 0, -1 / 3!, 0, 1 / 5!, 0, -1 / 7!, \ldots]$$

i.e., the corresponding function for *sin* is

sina (2 ∗ *n*)   = 0
sina (2 ∗ *n* + 1) = (−1)^*n* / (2 ∗ *n* + 1) !

This can be proven from the definitions of *cosx* and *sinx*.

Euler's formula and periodic functions   From this we obtain *Euler's formula*:

$exp\ (i * x) = cos\ x + i * sin\ x$

One thing which comes out of Euler's formula is the fact that the exponential is a *periodic function* along the imaginary axis. A function $f : A \to B$ is said to be periodic if there exists $T \in A$ such that

$f\ x = f\ (x + T)$ -- $\forall x \in A$

Therefore, for this definition to make sense, we need addition on A; in fact we normally assume at least *AddGroup A*.

Since *sin* and *cos* are periodic, with period $\tau = 2 * \pi$, we have, using the standard notation $a + i * b$ for some $z = a :+ b$:

$exp\ (z + i * \tau)$	$= \{\text{- Def. of }z\text{ -}\}$
$exp\ ((a + i * b) + i * \tau)$	$= \{\text{- Assoc. + distrib. -}\}$
$exp\ (a + i * (b + \tau))$	$= \{\text{- }H_2(exp, (+), (*))\text{ -}\}$
$exp\ a * exp\ (i * (b + \tau))$	$= \{\text{- Euler's formula -}\}$
$exp\ a * (cos\ (b + \tau) + i * sin\ (b + \tau))$	$= \{\text{- }cos\text{ and }sin\text{ are }\tau\text{-periodic -}\}$
$exp\ a * (cos\ b + i * sin\ b)$	$= \{\text{- Euler's formula -}\}$
$exp\ a * exp\ (i * b)$	$= \{\text{- }exp\text{ is a homomorphism -}\}$
$exp\ (a + i * b)$	$= \{\text{- Def. of }z\text{ -}\}$
$exp\ z$	

Thus, we see that *exp* is periodic, because $exp\ z = exp\ (z + T)$ with $T = i * \tau$, for all z.

Taylor meets transcendental functions   In Section 6.5 we saw that we could make power series instances for *Transcendental*, including *exp*, *sin*, and *cos*. We can do the same for the Taylor series representation: $[f, f', f'', \ldots]$. With this representation we have very easy implementations of derivative, integral, and value at zero:

```
derivT :: Taylor a → Taylor a
derivT = tail
integT :: a → Taylor a → Taylor a
integT = (:)
val :: Additive a ⇒ Taylor a → a
val (a: _) = a
val _      = zero
```

We can borrow the same derivation of the differential equations defining *expT*, *sinT*, and *cosT* recursively, just replacing *integ* with *integT*, etc.

expT, *sinT*, *cosT* :: *Transcendental a* ⇒ *Taylor a* → *Taylor a*
expT as = *integT* (*exp* (*val as*)) (*expT as* ∗ *derivT as*)
sinT as = *integT* (*sin* (*val as*)) (*cosT as* ∗ *derivT as*)
cosT as = *integT* (*cos* (*val as*)) (−*sinT as* ∗ *derivT as*)

With this in place we can assemble the instance declaration:

instance *Transcendental a* ⇒ *Transcendental* (*Taylor a*) **where**
$\pi = [\pi]$; *exp* = *expT*; *sin* = *sinT*; *cos* = *cosT*

Only one instance remains: we have not yet defined *recip* for Taylor series. Here we can use the easy access to the derivative as a help in the implementation. We want to compute g == *recip f* which is specified by $g * f$ == *one*. We can write f == *head f* : *tail f* == $f_0 : f'$ where f' :: *Taylor a*. If we compute the derivative of the specification we get *derivT* $(g * f)$ == *derivT one* which simplifies to $g' * f + g * f'$ == 0. From this we can extract g' == $-g * f' / f$ == $-f' * g * g$ because dividing by f is the same as multiplying by *recip f* == g. Thus we can implement the last remaining instance:

instance *Field a* ⇒ *MulGroup* (*Taylor a*) **where**
    recip = *recipStream*
recipStream :: *Field a* ⇒ *Taylor a* → *Taylor a*
recipStream [ ] = *error* "recipStream: divByZero"
recipStream $(f_0 : f') = g$
    where g = $g_0 : g'$
            g_0 = *recip* f_0
            g' = *negate* $(f' * g * g)$

Just to check we can now compute the first few terms in *expT x* : *Taylor* ℝ:

expTx :: *Taylor* ℝ
expTx = *expT* [0, 1]
testExpT = *take* 5 *expTx* == [1.0, 1.0, 1.0, 1.0, 1.0]

All derivatives are one, as expected from *exp* 0 = 1 and exp' = *exp*.

We can also combine this with the complex number example above:

expix3 :: *Taylor* (*Complex Double*)
expix3 = *exp* (*i* ∗ *x*)   **where** $i = [0 :+ 1]$; $x = [0, 1]$
ex2alt3 = *take* 10 *expix3*
testex2alt3 = *ex2alt3* == *take* 10 (*concat* (*repeat* [1, *i*, −1, −*i*]))

cosxisinx3 :: *Taylor* (*Complex Double*)
cosxisinx3 = *cos x* + *i* ∗ *sin x* **where** *i* = [0 :+ 1]; *x* = [0, 1]
ex2′alt3 = *take* 10 *cosxisinx3*

As with *expTx*, the coefficients in these series are also really simple: the four values [1, *i*, −1, −*i*] just repeat forever.

8.2   The Laplace transform

This material was inspired by Quinn and Rai [2008], which is highly recommended reading. In this section we typeset multiplication as (·) to avoid visual confusion with the convolution operator (⊛).

Consider the differential equation

$$f'' \; x - 3 \cdot f' \; x + 2 \cdot f \; x = exp \; (3 \cdot x), f \; 0 = 1, f' \; 0 = 0$$

We can solve such equations with the machinery of power series:

fs :: *Field a* ⇒ *PowerSeries a*
fs = *integ* 1 *fs′*
   where *fs′* = *integ* 0 *fs″*
           fs″ = *exp3x* + 3 · *fs′* − 2 · *fs*
exp3x :: *Field a* ⇒ *PowerSeries a*
exp3x = *compScale* 3 *expx*

We have done this by "zooming in" on the function *f* and representing it by a power series, $f \; x = \Sigma \; a_n \cdot x\hat{\ }n$. This allows us to reduce the problem of finding a function $f : \mathbb{R} \to \mathbb{R}$ to that of finding a list of coefficients a_n, or equivalently a function $a : \mathbb{N} \to \mathbb{R}$. Or even, if one wants an approximation only, finding a list of sufficiently many *a*-values for a good approximation.

Still, recursive equations are not always easy to solve (especially without a computer), so it's worth looking for alternatives.

When "zooming in" we go from *f* to *a*, but we can also look at it in the other direction: we have "zoomed out" from *a* to *f* via an infinite series:

$$a : \mathbb{N} \to \mathbb{R} \xrightarrow{\;\Sigma a_n \cdot x^n\;} f : \mathbb{R} \to \mathbb{R}$$

We would like to go one step further

$$a : \mathbb{N} \to \mathbb{R} \xrightarrow{\;\Sigma a_n \cdot x^n\;} f : \mathbb{R} \to \mathbb{R} \xrightarrow{\;??\;} F{:}?$$

That is, we are looking for a transformation from f to some F in a way which resembles the transformation from a to f. The analogue of "sum of an infinite series" for a continuous function is an integral:

$$a : \mathbb{N} \to \mathbb{R} \xrightarrow{\sum a_n \cdot x^n} f : \mathbb{R} \to \mathbb{R} \xrightarrow{\int (f\, t) \cdot x^t\, dt} F:?$$

We note that, for the integral $\int_0^\infty (f\, t) \cdot x^t\, dt$ to converge for a larger class of functions (say, bounded functions[1]), we have to limit ourselves to $|x| < 1$. Both this condition and the integral make sense for $x \in \mathbb{C}$, so we could take

$$a : \mathbb{N} \to \mathbb{R} \xrightarrow{\sum a_n \cdot x^n} f : \mathbb{R} \to \mathbb{R} \xrightarrow{\int (f\, t) \cdot x^t\, dt} F : \{z \mid |z| < 1\} \to \mathbb{C}$$

but let us stick to $\mathbb{R}$ for now.

Writing, somewhat optimistically

$$\mathscr{L} f\, x = \int_0^\infty (f\, t) \cdot x\hat{\,}t\, dt$$

we can ask ourselves what $\mathscr{L} f'$ looks like. After all, we want to solve *differential* equations by "zooming out". We have

$$\mathscr{L} f'\, x = \int_0^\infty (f'\, t) \cdot x\hat{\,}t\, dt$$

Remember that $D\, (f \cdot g) = D f \cdot g + f \cdot D\, g$, which we use with $g\, t = x\hat{\,}t$ so that $D\, g\, t = \log x \cdot x\hat{\,}t$ (note that t is the variable here, not x).

$\mathscr{L} f'\, x$	$= \{\text{- Def. of } \mathscr{L} \text{ -}\}$
$\int_0^\infty (D f\, t) \cdot x\hat{\,}t\, dt$	$= \{\text{- Derivative of product -}\}$
$\int_0^\infty (D\, (f\, t \cdot x\hat{\,}t)) - f\, t \cdot \log x \cdot x\hat{\,}t\, dt$	$= \{\text{- Linearity of } \int \text{ -}\}$
$\int_0^\infty (D\, (f\, t \cdot x\hat{\,}t))\, dt - \log x \cdot \int_0^\infty f\, t \cdot x\hat{\,}t\, dt$	$= \{\text{- Def. of integral to } \infty. \text{ -}\}$
$\lim_{t \to \infty} (f\, t \cdot x\hat{\,}t) - (f\, 0 \cdot x\hat{\,}0)$	
$\quad - \log x \cdot \int_0^\infty f\, t \cdot x\hat{\,}t\, dt$	$= \{\text{- } |x| < 1 \text{ -}\}$
$-f\, 0 - \log x \cdot \int_0^\infty f\, t \cdot x\hat{\,}t\, dt$	$= \{\text{- Def. of } \mathscr{L} \text{ -}\}$
$-f\, 0 - \log x \cdot \mathscr{L} f\, x$	

The factor $\log x$ is somewhat awkward. Let us therefore return to the definition of $\mathscr{L}$ and operate a change of variables. First some rewriting:

$$\mathscr{L} f\, x = \int_0^\infty (f\, t) \cdot x\hat{\,}t\, dt \qquad \Leftrightarrow \{\text{- } x = \exp\, (\log x) \text{ -}\}$$

$$\mathscr{L} f\, x = \int_0^\infty (f\, t) \cdot (\exp\, (\log x))\hat{\,}t\, dt \Leftrightarrow \{\text{- } (a\hat{\,}b)\hat{\,}c = a\hat{\,}(b \cdot c) \text{ -}\}$$

---

[1] A function is bounded if there exists a bound B such that for all x, $|f\, x| \leqslant B$.

$$\mathscr{L}\,f\,x = \int_0^\infty (f\,t) \cdot exp\,(log\,x \cdot t)\,dt$$

Since $log\,x < 0$ for $|x| < 1$, we make the substitution $-s = log\,x$. The condition $|x| < 1$ becomes $s > 0$ (or, in $\mathbb{C}$, $real\,s > 0$), and we have

$$\mathscr{L}\,f\,s = \int_0^\infty (f\,t) \cdot exp\,(-s \cdot t)\,dt$$

This is the definition of the Laplace transform of the function f. Going back to the problem of computing $\mathscr{L}\,f'$, we now have

$$\mathscr{L}\,f'\,s \qquad\qquad = \{\text{- The computation above with } s = -log\,x. \text{-}\}$$
$$-f\,0 + s \cdot \mathscr{L}\,f\,s$$

We have obtained

$$\mathscr{L}\,f'\,s = s \cdot \mathscr{L}\,f\,s - f\,0 \quad \text{-- The "Laplace-D" law}$$

From this, we can deduce

$$\mathscr{L}\,f''\,s \qquad\qquad\qquad = \{\text{- Laplace-D for } f' \text{-}\}$$
$$s \cdot \mathscr{L}\,f'\,s - f'\,0 \qquad\quad = \{\text{- Laplace-D for } f \text{-}\}$$
$$s \cdot (s \cdot \mathscr{L}\,f\,s - f\,0) - f'\,0 = \{\text{- Simplification -}\}$$
$$s\hat{}2 \cdot \mathscr{L}\,f\,s - s \cdot f\,0 - f'\,0$$

Exercise 8.1: what is the general formula for $\mathscr{L}\,f^{(k)}s$?

Returning to our differential equation, we have

$$f''\,x - 3 \cdot f'\,x + 2 \cdot f\,x = exp\,(3 \cdot x), f\,0 = 1, f'\,0 = 0$$
$$\Leftrightarrow \{\text{- point-free form -}\}$$
$$f'' - 3 \cdot f' + 2 \cdot f = exp \circ (3\cdot), f\,0 = 1, f'\,0 = 0$$
$$\Rightarrow \{\text{- applying } \mathscr{L} \text{ to both sides -}\}$$
$$\mathscr{L}\,(f'' - 3 \cdot f' + 2 \cdot f) = \mathscr{L}\,(exp \circ (3\cdot)), f\,0 = 1, f'\,0 = 0 \quad \text{-- Eq. (1)}$$

Remark: Note that this is a necessary condition, but not a sufficient one. The Laplace transform is not injective. For one thing, it does not take into account the behaviour of f for negative arguments. Because of this, we often assume that the domain of definition for functions to which we apply the Laplace transform is $\mathbb{R}_{\geq 0}$. For another, it is known that changing the values of f for a countable number of its arguments does not change the value of the integral.

According to the definition of $\mathscr{L}$ and because of the linearity of the integral, we have that, for any f and g for which the transformation is defined, and for any constants α and β

$$\mathscr{L} \left(\alpha \triangleleft f + \beta \triangleleft g \right) = \alpha \triangleleft \mathscr{L} f + \beta \triangleleft \mathscr{L} g$$

Note that this is an equality between functions. Indeed, recalling Chapter 7, in particular Section 7.4.1, we are working here with the vector space of functions (f and g are elements of it). Let us call the space of real functions $V = X \to \mathbb{R}$ (for some X) so that we have $f, g : V$. And let us call the space of functions that $\mathscr{L}$ returns $W = S \to \mathbb{C}$ for some suitable type S. The operator ($\triangleleft$) refers to scaling in a vector space — here scaling functions in V on the LHS and functions in W on the RHS. The above equation says that $\mathscr{L}$ is a linear transformation from V to W.

Applying this linearity property to the left-hand side of (1), we have for any s:

$$\mathscr{L} \left(f'' - 3 \cdot f' + 2 \cdot f \right) s$$
$$= \{ - \mathscr{L} \text{ is linear } - \}$$
$$\mathscr{L} f'' \, s - 3 \cdot \mathscr{L} f' \, s + 2 \cdot \mathscr{L} f \, s$$
$$= \{ - \text{ re-writing } \mathscr{L} f'' \text{ and } \mathscr{L} f' \text{ in terms of } \mathscr{L} f - \}$$
$$s\char`^2 \cdot \mathscr{L} f \, s - s \cdot f \, 0 - f' \, 0 - 3 \cdot (s \cdot \mathscr{L} f \, s - f \, 0) + 2 \cdot \mathscr{L} f \, s$$
$$= \{ - f \, 0 = 1, f' \, 0 = 0 - \}$$
$$(s\char`^2 - 3 \cdot s + 2) \cdot \mathscr{L} f \, s - s + 3$$
$$= \{ - \text{ Factoring } - \}$$
$$(s - 1) \cdot (s - 2) \cdot \mathscr{L} f \, s - s + 3$$

For the right-hand side, we apply the definition:

$$\mathscr{L} \left(exp \circ (3 \cdot) \right) s \qquad\qquad = \{ - \text{ Def. of } \mathscr{L} - \}$$
$$\int_0^\infty exp \, (3 \cdot t) \cdot exp \, (-s \cdot t) \, dt \quad =$$
$$\int_0^\infty exp \, ((3 - s) \cdot t)) \, dt \qquad =$$
$$lim_{t \to \infty} \frac{exp \, ((3-s) \cdot t)}{3 - s} - \frac{exp \, ((3-s) \cdot 0)}{3 - s} = \{ - \text{ for } s > 3 - \}$$
$$\frac{1}{s - 3}$$

Therefore, we have, writing F for $\mathscr{L} f$:

$$(s - 1) \cdot (s - 2) \cdot F \, s - s + 3 = \frac{1}{s - 3}$$

and therefore, by solving for $F \, s$ we get

$$F \, s = \frac{\frac{1}{s-3} + s - 3}{(s - 1) \cdot (s - 2)} = \frac{10 - 6 \cdot s + s\char`^2}{(s - 1) \cdot (s - 2) \cdot (s - 3)}$$

We now have the problem of "recovering" the function f from its Laplace transform. The standard approach is to use the linearity of $\mathscr{L}$ to write F as a sum of functions with known inverse transforms. We know one such function:

$$\lambda t \to exp\ (\alpha \cdot t)\{\text{- is the inverse Laplace transform of -}\}\ \lambda s \to \tfrac{1}{s-\alpha}$$

In fact, in our case, this is all we need.

The idea is to write $F\ s$ as a sum of three fractions with denominators $s-1$, $s-2$, and $s-3$ respectively, i.e., to find A, B, and C such that

$$\tfrac{A}{s-1} + \tfrac{B}{s-2} + \tfrac{C}{s-3} = \tfrac{10-6 \cdot s + s2}{(s-1) \cdot (s-2) \cdot (s-3)}$$
$$\Rightarrow \{\text{- Multiply both sides by } (s-1) \cdot (s-2) \cdot (s-3) \text{ -}\}$$
$$A \cdot (s-2) \cdot (s-3) + B \cdot (s-1) \cdot (s-3) + C \cdot (s-1) \cdot (s-2)$$
$$= 10 - 6 \cdot s + s\hat{}2 \quad \text{-- (2)}$$

We need this equality (2) to hold for values $s > 3$. A *sufficient* condition for this is for (2) to hold for *all* s. A *necessary* condition for this is for (2) to hold for the specific values 1, 2, and 3.

$$\text{For } s = 1 : A \cdot (-1) \cdot (-2) = 10 - 6 + 1 \ \Rightarrow A = \tfrac{5}{2}$$
$$\text{For } s = 2 : B \cdot 1 \cdot (-1) \quad\ \ = 10 - 12 + 4 \Rightarrow B = -2$$
$$\text{For } s = 3 : C \cdot 2 \cdot 1 \qquad = 10 - 18 + 9 \Rightarrow C = \tfrac{1}{2}$$

It is now easy to check that, with these values, (2) does indeed hold, and therefore that we have

$$F\ s = \tfrac{5}{2} \cdot \tfrac{1}{s-1} - 2 \cdot \tfrac{1}{s-2} + \tfrac{1}{2} \cdot \tfrac{1}{s-3}$$

The inverse transform is now easy:

$$f\ t = \tfrac{5}{2} \cdot exp\ t - 2 \cdot exp\ (2 \cdot t) + \tfrac{1}{2} \cdot exp\ (3 \cdot t)$$

Our mix of necessary and sufficient conditions makes it necessary to check that we have, indeed, a solution for the differential equation. To do this we compute the first and second derivatives of f:

$$f'\ t = \tfrac{5}{2} \cdot exp\ t - 4 \cdot exp\ (2 \cdot t) + \tfrac{3}{2} \cdot exp\ (3 \cdot t)$$
$$f''\ t = \tfrac{5}{2} \cdot exp\ t - 8 \cdot exp\ (2 \cdot t) + \tfrac{9}{2} \cdot exp\ (3 \cdot t)$$

We then check the main equation:

LHS

$=$

$f'' x - 3 \cdot f' x + 2 \cdot f x$

$= \{$- Fill in the computed definitions of f, f', and f''. -$\}$

$\qquad \frac{5}{2} \cdot exp\ x - 8 \cdot exp\ (2 \cdot x) + \frac{9}{2} \cdot exp\ (3 \cdot x)$

$\qquad - 3 \cdot (\frac{5}{2} \cdot exp\ x - 4 \cdot exp\ (2 \cdot x) + \frac{3}{2} \cdot exp\ (3 \cdot x))$

$\qquad + 2 \cdot (\frac{5}{2} \cdot exp\ x - 2 \cdot exp\ (2 \cdot x) + \frac{1}{2} \cdot exp\ (3 \cdot x))$

$= \{$- Collect common terms -$\}$

$\qquad (1 - 3 + 2) \cdot \qquad\qquad\qquad \cdot \frac{5}{2} \cdot exp\ x$

$\qquad + (-8 - 3 \cdot (-4) + 2 \cdot (-2)) \cdot \quad exp\ (2 \cdot x)$

$\qquad + (9 - 3 \cdot 3 + 2 \cdot 1) \qquad\quad \cdot \frac{1}{2} \cdot exp\ (3 \cdot x)$

$= \{$- Arithmetics -$\}$

$\qquad \frac{2}{2} \cdot exp\ (3 \cdot x)$

$=$

RHS

Finally, we check the initial conditions: $f\ 0 = 1$ and $f'\ 0 = 0$. Here we use that $exp\ (\alpha \cdot 0) = exp\ 0 = 1$ for all α:

$$f\ 0 = \frac{5}{2} \cdot exp\ 0 - 2 \cdot exp\ (2 \cdot 0) + \frac{1}{2} \cdot exp\ (3 \cdot 0)$$
$$= \frac{5}{2} - 2 + \frac{1}{2}$$
$$= 1$$

$$f'\ 0 = \frac{5}{2} \cdot exp\ 0 - 4 \cdot exp\ (2 \cdot 0) + \frac{3}{2} \cdot exp\ (3 \cdot 0)$$
$$= \frac{5}{2} - 4 + \frac{3}{2}$$
$$= 0$$

Thus we can conclude that our f does indeed solve the differential equation. The checking may seem overly pedantic, but when solving these equations by hand it is often these last checks which help catching mistakes along the way.

8.2.1 Some standard Laplace transforms

Now when we have tools to calculate solutions to ODEs we can use this to compute some Laplace transforms for common functions specified by ODEs, like *exp*, *sin*, and *cos*.

Deriving $\mathscr{L}\ exp$: The easiest one is *exp* which is the unique solution to the ODE $e' = e$, $e\ 0 = 1$. Here we can use the "Laplace-D" law: $\mathscr{L}\ e'\ s = s \cdot \mathscr{L}\ e\ s - e\ 0$ in combination with the initial condition and the application of $\mathscr{L}$ to our characterising ODE $\mathscr{L}\ e'\ s = \mathscr{L}\ e\ s$ to eliminate $\mathscr{L}\ e'\ s$. We get

$\mathscr{L}$ e s = $s \cdot \mathscr{L}$ e s − 1 which is now an equation we can solve for $\mathscr{L}$ e s (abbreviated to E s). First collect the terms, $(s − 1) \cdot E$ s = 1, then divide by $s − 1$ to get E s = 1 / $(s − 1)$. Thus $\mathscr{L}$ exp s = 1 / $(s − 1)$.

It is instructive to see what happens when we make minor changes to the ODE, for example $f' = \alpha \cdot f, f$ 0 = 1 for some constant α. The derivation is very similar but we get the equation $\alpha \cdot \mathscr{L}$ f s = $s \cdot \mathscr{L}$ f s − 1 to solve, which gives us $\mathscr{L}$ f s = 1 / $(s − \alpha)$. It is reassuring to see that we recover exp if we let α = 1. But what function is f? We know that f x = $A \cdot exp$ $(\alpha \cdot x)$ has derivative f' x = $\alpha \cdot A \cdot exp$ $(\alpha \cdot x)$ = $\alpha \cdot f$ x, and with A = 1 we satisfy the initial condition f 0 = 1. Thus $\mathscr{L}$ $(\lambda x \rightarrow exp$ $(\alpha \cdot x))$ s = 1 / $(s − \alpha)$.

From this we can also see that the ODE $f' = \alpha \cdot f, f$ 0 = A has the transform A / $(s − \alpha)$.

Deriving $\mathscr{L}$ sin and $\mathscr{L}$ cos: With the same method we can compute the transforms of sin and cos. First the ODEs: sin and cos are the solutions to the coupled equations $si' = co, si$ 0 = 0, $co' = −si, co$ 0 = 1. We apply Laplace to both ODEs (and linearity in the second):

$$\mathscr{L}\ si'\ s = \mathscr{L}\ co\ \ \ \ s$$
$$\mathscr{L}\ co'\ s = \mathscr{L}\ (−si)\ s = −\mathscr{L}\ si\ s$$

As for exp we can use the Laplace-D-law to simplify the derivatives:

$$\mathscr{L}\ si'\ s = s \cdot \mathscr{L}\ si\ s − si\ 0 = s \cdot \mathscr{L}\ si\ s$$
$$\mathscr{L}\ co'\ s = s \cdot \mathscr{L}\ co\ s − co\ 0 = s \cdot \mathscr{L}\ co\ s − 1$$

We combine our equations to eliminate si' and co':

$$\mathscr{L}\ co\ s = s \cdot \mathscr{L}\ si\ s$$
$$−\ \mathscr{L}\ si\ s = s \cdot \mathscr{L}\ co\ s − 1$$

and introduce names for the transformed functions: $S = \mathscr{L}$ si and $C = \mathscr{L}$ co:

$$C\ s = s \cdot S\ s$$
$$−\ S\ s = s \cdot C\ s − 1$$

Use the first to eliminate C s in the second to get $−S$ s = $s\hat{}2 \cdot S$ s − 1 which means 1 = $(s\hat{}2 + 1) \cdot S$ s and finally

$$S\ s = \ \ \ \ \ \ \ \ \ \ \ \ 1\ /\ (s\hat{}2 + 1)$$
$$C\ s = s \cdot S\ s = s\ /\ (s\hat{}2 + 1)$$

8.3 Laplace and other transforms

To sum up, we have defined the Laplace transform and shown that it can be used to solve differential equations. It can be seen as a continuous version of the transform between the infinite sequence of coeeficients $a : \mathbb{N} \to \mathbb{R}$ and the functions behind formal power series.

The Laplace transform is a close relative of the Fourier transform. Both transforms are used to express functions as a sum of "complex frequencies", but Laplace allows a wider range of functions to be transformed. A nice overview and comparison is B. Berndtsson's "Fourier and Laplace Transforms"[2]. Fourier analysis is a common tool in courses on Transforms, Signals and Systems.

The Fourier transform can be seen as a generalisation of the Fourier series presented in Section 7.4.3. The Fourier series is a way of expressing functions on a *closed* interval (or, equivalently, periodic functions on the real line) as a linear combination of dicrete frequency components rather than as a function of time. The Fourier transform (like the Laplace transform) also handles non-periodic functions, and the result is a continuous linear combination of frequency components (an integral rather than a sum).

---

[2]Available from `http://www.math.chalmers.se/Math/Grundutb/CTH/mve025/1516/Dokument/F-analys.pdf`.

8.4   Exercises

Exercise 8.1. Starting from the "Laplace-D" law

$$\mathscr{L} f' \, s = s \cdot \mathscr{L} f \, s - f \, 0$$

Derive a general formula for $\mathscr{L} f^{(k)} s$.

Exercise 8.2. Find the Laplace transforms of the following functions:

1. $\lambda t \to 3 \cdot e^{5 \cdot t}$

2. $\lambda t \to e^{\alpha \cdot t} - \beta$

3. $\lambda t \to e^{t + \frac{\pi}{6}}$

Exercise 8.3.

1. Show that:
   - $\sin t = \frac{1}{2 \cdot i} \cdot (e^{i \cdot t} - e^{-i \cdot t})$
   - $\cos t = \frac{1}{2} \cdot (e^{i \cdot t} + e^{-i \cdot t})$

2. Find the Laplace transforms $\mathscr{L} \sin$ and $\mathscr{L} \cos$ using the transform for the exponentials and the result from Item 1.

Exercise 8.4. *From exam 2016-03-15*

Consider the following differential equation:

$$f'' \, t - 2 \cdot f' \, t + f \, t = e^{2 \cdot t}, \quad f \, 0 = 2, \quad f' \, 0 = 3$$

Solve the equation using the Laplace transform. You should need only one formula (and linearity):

$$\mathcal{L} \left(\lambda t. \, e^{\alpha \cdot t} \right) s = 1/(s - \alpha)$$

Exercise 8.5. *From exam 2016-08-23*

Consider the following differential equation:

$$f'' \, t - 5 \cdot f' \, t + 6 \cdot f \, t = e^{t}, \quad f \, 0 = 1, \quad f' \, 0 = 4$$

Solve the equation using the Laplace transform. You should need only one formula (and linearity):

$$\mathcal{L} \left(\lambda t. \, e^{\alpha \cdot t} \right) s = 1/(s - \alpha)$$

Exercise 8.6. *From exam 2016-Practice*

Consider the following differential equation:

$$f''t - 2 \cdot f't + ft - 2 = 3 \cdot e^{2 \cdot t}, \quad f0 = 5, \quad f'0 = 6$$

Solve the equation using the Laplace transform. You should need only one formula (and linearity):

$$\mathcal{L}\left(\lambda t. e^{\alpha \cdot t}\right)s = 1/(s - \alpha)$$

Exercise 8.7. *From exam 2017-03-14*

Consider the following differential equation:

$$f''t + 4 \cdot ft = 6 \cdot \cos t, \quad f0 = 0, \quad f'0 = 0$$

Solve the equation using the Laplace transform. You should need only two formulas (and linearity):

$$\mathcal{L}\left(\lambda t. e^{\alpha \cdot t}\right)s = 1/(s - \alpha)$$

$$2 \cdot \cos t = e^{i \cdot t} + e^{-i \cdot t}$$

Exercise 8.8. *From exam 2017-08-22*

Consider the following differential equation:

$$f''t - 3\sqrt{2} \cdot f't + 4 \cdot ft = 0, \quad f0 = 2, \quad f'0 = 3\sqrt{2}$$

Solve the equation using the Laplace transform. You should need only one formula (and linearity):

$$\mathcal{L}\left(\lambda t. e^{\alpha \cdot t}\right)s = 1/(s - \alpha)$$

Chapter 9

Probability Theory

We have by now acquired a firm grip on DSL notions and several mathematical domains. In this chapter, will apply the DSL methodology once more, and now to the domain of probability theory. By building a DSL from scratch, we will not only clarify notations for (conditional) probabilities, such as $P(A \mid B)$, but we will also be able to describe and reason about problems such as those in the following list. For them we will even be able to compute the probabilities involved by evaluating the DSL expressions.

1. Assume you throw two 6-faced dice, what is the probability that the product is greater than 10 if their sum is greater than 7?

2. Suppose that a test for using a particular drug is 99% sensitive and 99% specific. That is, the test will produce 99% true positive results for drug users and 99% true negative results for non-drug users. Suppose that 0.5% of people are users of the drug. What is the probability that a randomly selected individual with a positive test is a drug user? (Example found in Wikipedia article on Bayes' theorem, 2019-03-01.)

3. Suppose you are on Monty Hall's *Let's Make a Deal!* You are given the choice of three doors, behind one door is a car, the others, goats. You pick a door, say 1, Monty opens another door, say 3, which has a goat. Monty says to you "Do you want to pick door 2?" Assuming that the car is more desirable than the goat, is it to your advantage to switch your choice of doors?

Our method will be to:

- Describe the space of possible situations, or outcomes.

- Define the events whose probabilities we will consider.

- Evaluate such probabilities.

9.1 Sample spaces

Generally, textbook problems involving probability involve the description of some scenario or experiment, with an explicit uncertainty, including the outcome of certain measurements.[1] Then the reader is asked to compute the probability of some event.

It is common to refer to a sample space by the labels S, Ω, or U, but in this chapter we will define many such spaces, and therefore we will use descriptive names instead. While the concept of a space of events underpins modern understandings of probability theory, textbooks sometimes give a couple of examples involving coin tosses and promptly forget the concept of sample space in the body of the text. Here we will instead develop this concept using our DSL methodology. Once this is done, we will be able to see that an accurate model of the sample space is an essential tool to solve probability problems.

Our first task is to describe the possible structure of sample spaces, and model them as a DSL. To this end we will use a data type to represent spaces. This type is indexed by the underlying Haskell type of possible outcomes. Hence *Space* maps this underlying type to another type:

$$Space :: Type \rightarrow Type$$

We then carry on and define the constructions which inhabit the above type.

Finite space In Example 1., we consider dice with 6 faces. For this we define a constructor *Finite* embedding a list of possible outcomes into a space:

$$Finite :: [a] \rightarrow Space\ a$$

---

[1]Depending on the context, we use the word "situation" or "outcome" for the same mathematical objects. The word "outcome" evokes some experiment, explicitly performed; and the "outcome" is the situation after the experiment is over. When we use the word "situation" there is not necessarily an explicit experiment, but something happens according to a specific scenario. In this case we call the "situation" the state of affairs at end of the scenario in question.

The scenario (or "experiment") corresponding to throwing a general n-sided die is then represented by the space *die n*:

> *die* :: *Int* $\rightarrow$ *Space Int*
> *die n* = *Finite* $[1 . . n]$
> *d6* = *die* 6

In particular the space *point x* is the space with a single point — only trivial probabilities (zero or one) are involved here:

> *point* :: *a* $\rightarrow$ *Space a*
> *point x* = *Finite* $[x]$

Scaling space   If the die is well-balanced, then all cases have the same probability (or probability mass) in the space, and this is what we have modelled above. But this is not always the case. Hence we need a way to represent such imbalances. We use the following combinator:[2]

> *Factor* :: $\mathbb{R}$ $\rightarrow$ *Space* ()

Its underlying type is the unit type (), but its mass (or density) is given by a real number.

On its own, *Factor* may appear useless, but we can setup the *Space* type so that this mass or density can depend on (previously introduced) spaces.

Product of spaces   As a first instance of a dependency, we introduce the product of two spaces, as follows:

> *prod* :: *Space a* $\rightarrow$ *Space b* $\rightarrow$ *Space* (a, b)

For example, the "experiment" of throwing two 6-faced dice is represented as follows:[3]

> *twoDice* :: *Space* (Int, Int)
> *twoDice* = *prod d6 d6*

---

[2]This is in fact *scaling*, as defined in Chapter 7. Indeed, there is a vector space of measurable spaces, where each space is one vector. However, we choose not to use this terminology, because we are generally not interested in the vector space structure of probability spaces. There is also potential for confusion, because *Factor* does not scale the *points* in the space. What it does is to scale the *probability mass* associated with each such points.

[3]The use of a pair corresponds to the fact that the two dice can be identified individually.

But let's say now that we know that the sum is greater than 7. We can then define the following parametric space, whose single point has mass 1 if the condition is satisfied and 0 otherwise. (This space is trivial in the sense that no uncertainty is involved.)

> *sumAbove7* :: $(Int, Int) \rightarrow$ *Space* ()
> *sumAbove7* $(x, y) =$ *Factor* (**if** $x + y > 7$ **then** 1 **else** 0)

We now want to take the product of the *twoDice* space and the *sumAbove7* space; but the issue is that *sumAbove7 depends* on the outcome of *twoDice*. To support this dependency we need a generalisation of the product which we call "Sigma" (Σ) because of its similarity of structure with "big sum".

> Σ :: *Space a* $\rightarrow$ $(a \rightarrow Space\ b)$ $\rightarrow$ *Space* (a, b)

Hence:

> *problem1* :: *Space* $((Int, Int), ())$
> *problem1* $=$ Σ *twoDice sumAbove7*

The values of the dice are the same as in *twoDice*, but the density of any sum less than 7 is brought down to zero.

We can check that the product of spaces is a special case of Σ:

> *prod a b* $=$ Σ *a* (*const b*)

Projections   In the end we may not be interested in all values and hide some of them. For this purpose we use the following combinator:

> *Project* :: $(a \rightarrow b) \rightarrow$ *Space a* $\rightarrow$ *Space b*

A typical use is *Project fst* :: *Space* $(a, b) \rightarrow$ *Space a*, ignoring the second component of a pair. We can make *Space* an instance of the *Functor* type class with *fmap* $=$ *Project*.

Real line   Before we continue, we may also add a way to represent real-valued spaces, which assign an even probability density across all reals.

> *RealLine* :: *Space* $\mathbb{R}$

To get other probability distributions over the reals, you can combine *RealLine* with *Factor* — see, for example, the normal distribution in Section 9.3.

Summary We have already completed the description of a DSL for spaces, whose abstract syntax is captured by the following datatype:

data *Space a* **where**
$\quad$ *Finite* $\quad$ :: $[a] \to$ *Space a*
$\quad$ *Factor* $\quad$ :: $\mathbb{R} \to$ *Space* $()$
$\quad$ Σ $\qquad$ :: *Space a* $\to$ $(a \to$ *Space b*$) \to$ *Space* (a,b)
$\quad$ *Project* $\;$ :: $(a \to b) \to$ *Space a* $\to$ *Space b*
$\quad$ *RealLine* :: *Space* $\mathbb{R}$

9.2 *Monad Interface

Seasoned functional programmers will be aware of monadic interfaces. For them, it may be useful to know that one can easily provide a monadic interface for spaces. The implementation is the following:

instance *Functor Space* **where**
$\quad$ *fmap* $\;$ = *Project*
instance *Applicative Space* **where**
$\quad$ *pure* $\;\;$ = *point*
$\quad$ $(<\!*\!>)$ = *Control.Monad.ap*
instance *Monad Space* **where**
$\quad$ $a \ggeq f$ = *Project snd* $(\Sigma\, a\, f)$

Exercise 9.1. Prove the functor and monad laws for the above definitions. (Use semantic equality, see Section 9.4.)

9.3 Distributions

So far we have defined several spaces, but we have not used them to compute any probability. We set out to do this in this section. In section 9.4 we will see how to compute the total mass (or *measure* :: *Space a* $\to \mathbb{R}$) of a space. But before that we will talk about another important notion in probability theory: that of a distribution. A distribution is a space whose *measure* is equal to 1.

$\quad$ *isDistribution* :: *Space a* $\to$ *Bool*
$\quad$ *isDistribution s* = *measure s* == 1

We may use the following type synonym to indicate distributions:

> **type** *Distr a* = *Space a*

Let us define a few useful distributions. First, we present the uniform distribution among a finite set of elements. It is essentially the same as the *Finite* space, but we scale it so that the total measure comes down to 1.

> *uniformDiscrete* :: $[a]$ → *Distr a*
> *uniformDiscrete xs* = *scale* $(1.0 \;/\; fromIntegral \; (length \; xs))$
> $\qquad\qquad\qquad\qquad (Finite \; xs)$

The distribution of the balanced die can then be represented as follows:

> *dieDistr* :: *Distr Integer*
> *dieDistr* = *uniformDiscrete* $[1 \mathinner{.\,.} 6]$

Scaling is a special case of the more general operation *marginaliseWith* which applies a factor to every point, and ignores the unit type from *Factor*. This operation is often called "marginalisation", in the jargon of Bayesian reasoning.

> *marginaliseWith* :: $(a → \mathbb{R})$ → *Space a* → *Space a*
> *marginaliseWith f s* = *Project fst* $(\Sigma \; s \; (\lambda x → Factor \; (f \; x)))$
>
> *scale* :: $\mathbb{R}$ → *Space a* → *Space a*
> *scale c* = *marginaliseWith* $(const \; c)$

If the scaling is all-or-nothing, we have the following version, which will be useful later.

> *observing* :: $(a → Bool)$ → *Space a* → *Space a*
> *observing f* = *marginaliseWith* $(indicator \circ f)$
>
> *indicator* :: *Bool* → $\mathbb{R}$
> *indicator True* = 1
> *indicator False* = 0

Another useful discrete distribution is the Bernoulli distribution of parameter p. It is a distribution whose value is *True* with probability p and *False* with probability $1 - p$. Hence it can be used to represent a biased coin toss.

> *bernoulli* :: $\mathbb{R}$ → *Distr Bool*
> *bernoulli p* = *marginaliseWith* $(\lambda b → $ **if** b **then** p **else** $1 - p)$
> $\qquad\qquad\qquad\qquad (Finite \; [False, True])$

Finally we can define the normal distribution with average μ and standard deviation σ.

$normal :: \mathbb{R} \rightarrow \mathbb{R} \rightarrow Distr\ \mathbb{R}$
$normal\ \mu\ \sigma = marginaliseWith\ (normalMass\ \mu\ \sigma)\ RealLine$
$normalMass :: Floating\ r \Rightarrow r \rightarrow r \rightarrow r \rightarrow r$
$normalMass\ \mu\ \sigma\ x = exp\ (-y\hat{\ }2\ /\ 2)\ /\ (\sigma * \sqrt{2 * \pi})$
    where $y = (x - \mu)\ /\ \sigma$

In scientific literature, distributions are sometimes called "random variables". However we consider this terminology to be misleading — random variables will be defined precisely later on.

We could try to define the probabilities or densities of possible values of a distribution, say $distDensityAt :: Space\ a \rightarrow (a \rightarrow \mathbb{R})$, and from there define the expected value (and other statistical moments), but we will take another route.

9.4 Semantics of spaces

First, we come back to general probability spaces without restriction on their *measure*: it does not need to be equal to one. We define a function *integrator*, which generalises the notions of weighted sum, and weighted integral. When encountering *Finite* spaces, we sum; when encountering *RealLine* we integrate. When encountering *Factor* we will adjust the weights. The integrator of a product (in general Σ) is the nested integration of spaces. The weight is given as a second parameter to *integrator*, as a function mapping elements of the space to a real value.

In code, we obtain the following:[4]

$integrator :: Space\ a \rightarrow (a \rightarrow \mathbb{R}) \rightarrow \mathbb{R}$
$integrator\ (Finite\ a) \quad g = bigsum\ a\ g$
$integrator\ (RealLine) \quad g = integral\ g$
$integrator\ (Factor\ f) \quad g = f * g\ ()$
$integrator\ (\Sigma\ a\ f) \quad g = integrator\ a \qquad \$\ \lambda x \rightarrow$
$\qquad\qquad\qquad\qquad\qquad integrator\ (f\ x)\ \$\ \lambda y \rightarrow$
$\qquad\qquad\qquad\qquad\qquad g\ (x, y)$
$integrator\ (Project\ p\ a)\ g = integrator\ a\ (g \circ p)$

In calculations we will often use the notation $\int \{s\}\ g$ for *integrator s g*. This shows that we are dealing with a generalisation of integration to more complicated domains (spaces). For simplicity we use $\mathbb{R}$ here, but the definitions would work for any field.

---

[4]You may want to come back to Section 1.7.2 to see how to deal with the integration (or summation) variable and what it means for the type of the integrator.

The above definition relies on the usual notions of sum (*bigsum*) and *integral*. We can define the sum of some terms, for finite lists, as follows:

$bigsum :: [a] \rightarrow (a \rightarrow \mathbb{R}) \rightarrow \mathbb{R}$
$bigsum\ xs\ f = sum\ (map\ f\ xs)$

We use also the definite integral over the whole real line. However, at the Haskell level, we will leave this concept undefined — thus whenever using real-valued spaces, our definitions are not usable for numerical computations, but for symbolic computations only. (If we had a syntax for the function to integrate, we could do more, but this would take us too far off track here.)

$integral :: (\mathbb{R} \rightarrow \mathbb{R}) \rightarrow \mathbb{R}$
$integral = undefined$

This semantics yields a notion of semantic equality for space, which we can define as follows:

$s_1 === s_2 = \forall\ g.\ integrator\ s_1\ g === integrator\ s_2\ g$

The simplest useful quantity that we can compute using the integrator is the measure of the space — its total "mass" or "volume". To compute the measure of a space, we can simply integrate the constant 1 (so only the mass of the space matters).

$measure :: Space\ a \rightarrow \mathbb{R}$
$measure\ d = integrator\ d\ (const\ 1)$

As a sanity check, we can compute the measure of a Bernoulli distribution (we use the $\ggg$ notation to indicate an expression being computed) and find that it is indeed 1.

    -- >>> measure (bernoulli 0.2)
    -- 1.0

The integration of *id* over a real-valued distribution yields its expected value.[5]

$expectedValueOfDistr :: Distr\ \mathbb{R} \rightarrow \mathbb{R}$
$expectedValueOfDistr\ d = integrator\ d\ id$

    -- >>> expectedValueOfDistr dieDistr
    -- 3.5

Exercise 9.2. Compute symbolically the expected value of the *bernoulli* distribution.

---

[5]We reserve the name *expectedValue* for the expected value of a random variable, defined later.

Properties of *integrator* We can use calculational reasoning to show some useful properties of spaces.

Lemma 9.1 (Linearity of *integrator*). If g is a linear function (addition or multiplication by a constant), then:

$$integrator\ s\ (g \circ f) == g\ (integrator\ s\ f)$$

(or, equivalently, *integrator* s $(\lambda x \to g\ (f\ x)) = g\ (integrator\ s\ f)$)

Proof. The proof proceeds by structural induction over s. The hypothesis of linearity is used in the base cases. Notably the linearity property of *Finite* and *RealLine* hinges on the linearity of sums and integrals.[6] The case of *Project* is immediate by definition. The case for Σ is proven as follows:

$\int \{\Sigma\ a\ f\}\ (g \circ h)$
$= \{\text{- By definition -}\}$
$\int \{a\}\ \$\ \lambda x \to \int \{f\ x\}\ \$\ \lambda y \to g\ (h\ (x,y))$
$= \{\text{- By induction -}\}$
$\int \{a\}\ \$\ \lambda x \to g\ (\int \{f\ x\}\ \$\ \lambda y \to h\ (x,y)$
$= \{\text{- By induction -}\}$
$g\ (\int \{a\}\ \$\ \lambda x \to (\int \{f\ x\}\ \$\ \lambda y \to h\ (x,y))$
$= \{\text{- By definition -}\}$
$g\ (\int \{\Sigma\ a\ f\}\ h)$

$\square$

Lemma 9.2 (Properties of *measure*).

- *measure* $(Finite\ xs) == length\ xs$

- *measure* $(prod\ s\ t) == measure\ s * measure\ t$.

- If s is a distribution and $f\ x$ is a distribution for every x, then $\Sigma\ s\ f$ is a distribution

- s is a distribution iff. *Project* $f\ s$ is a distribution.

The proof of the second item is as follows:

$measure\ (prod\ s\ t)$
$== \{\text{- By definition of } prod \text{ -}\}$
$measure\ (\Sigma\ s\ (const\ t))$
$== \{\text{- By definition of measure -}\}$

---

[6]Recall that integration is a linear operator in the space of functions (Section 7.4.1).

integrator (Σ *s* (*const t*)) (*const* 1)
== {- By definition of integrator for Σ -}
  integrator s $ \lambda x \rightarrow$ *integrator* (*const t x*) (*const* 1)
== {- By definition of measure, const -}
  integrator s $ \lambda x \rightarrow$ *measure t*
== {- By linearity of integrator -}
  measure t $*$ *integrator s* (*const* 1)
== {- By definition of measure -}
  measure t $*$ *measure s*

Exercise 9.3. Using the above lemmas, prove:

integrator (*bernoulli p*) *f* == *p* $*$ *f True* $+$ $(1 - p)$ $*$ *f False*

9.5   Random Variables

Even though we studied variables in detail (in Chapter 3), it is good to come
back to them for a moment before returning to *random* variables proper.

According to Wikipedia[7], a variable has a different meaning in computer science and in mathematics:

> Variable may refer to:
>
> - Variable (computer science): a symbolic name associated with
>   a value and whose associated value may be changed
> - Variable (mathematics): a symbol that represents a quantity
>   in a mathematical expression, as used in many sciences

At this stage of the book, we have a pretty good grip on variables in computer science. In particular, in Chapter 3, we have described a way to reduce
mathematical variables (position q and velocity v in Lagrangian mechanics)
to computer science variables. This was done by expressing variables as *functions* of a truly free variable, time (t). Time is a "computer science" variable in
this context: it can be substituted by any value without "side effects" on other
variables (unlike positions (q) and velocities (v)).

In this light, let us return to random variables. Wikipedia is not very helpful
here, so we turn to Grinstead and Snell [2003], who give the following definition:

---

[7]https://en.wikipedia.org/wiki/Variable, retrieved 2021-11-01.

> A random variable is simply an expression whose value is the outcome of a particular experiment.

This may be quite confusing at this stage. What are those expressions in our DSL? And where does the experiment influence the variable? Our answer is to use spaces to represent the "experiments" that Grinstead and Snell mention. More specifically, if $s : Space\ a$, then each possible situation at the end of the experiment is representable in the type a and the space will specify the mass of each of them (formally, via the integrator).

Then, a b-valued random variable (observed after an experiment represented by a space $s : Space\ a$) is a function f of type $a \rightarrow b$. Then $f\ x$ is the "expression" that Grinstead and Snell refer to. The (computer science) variable x is the outcome, and s represents the experiment — which is most often implicit in a random variable expressions as written in a mathematics book.

We can finally define the expected value (and other statistical moments) of a random variable.

In textbooks, one will often find the notation $E[t]$ for the expected value of a real-valued random variable t. As just mentioned, from our point of view, this notation can be confusing because it leaves the space of situations completely implicit. That is, it is not clear how t depends on the outcome of experiments.

With our DSL approach, we make this dependency completely explicit. For example, the expected value of a real-valued random variable takes the space of outcomes as its first argument:

> *expectedValue* :: $Space\ a \rightarrow (a \rightarrow \mathbb{R}) \rightarrow \mathbb{R}$
> *expectedValue* $s\ f = integrator\ s\ f\ /\ measure\ s$

For instance, we can use the above function to compute the expected value of the sum of two dice throws:

> *expect2D6* $= expectedValue\ twoDice\ (\lambda(x, y) \rightarrow fromIntegral\ (x + y))$

Exercise 9.4. Run the above code and check that you obtain the value 7.

Essentially, what the above definition of expected value does is to compute the weighted sum/integral of $f\ x$ for every point x in the space. And because s is a space (not necessarily a distribution), we must normalise the result by dividing by its measure.

Exercise 9.5. Define various statistical moments (variance, skew, curtosis, etc.)

9.6   Events and probability

In textbooks, one typically finds the notation $P(e)$ for the probability of an *event e*. Again, the space of situations s is implicit as well as the dependency between e and s.

Here, we define events as *Boolean-valued* random variables. Thus an event e can be defined as a Boolean-valued function $e : a \rightarrow Bool$ over a space $s : Space\ a$. Assuming that the space s accurately represents the relative mass of all possible situations, there are two ways to define the probability of e.

The first definition of the probability of e is the expected value of *indicator* $\circ$ *e*:

> *probability1* :: *Space a* $\rightarrow$ ($a \rightarrow Bool$) $\rightarrow \mathbb{R}$
> *probability1 d e* = *expectedValue d* (*indicator* $\circ$ *e*)

The second definition of probability is the ratio between the measure of the subspace where e holds, and the measure of the complete space.

> *probability2* :: *Space a* $\rightarrow$ ($a \rightarrow Bool$) $\rightarrow \mathbb{R}$
> *probability2 s e* = *measure* ($\Sigma\ s$ (*isTrue* $\circ$ *e*)) / *measure s*
>
> *isTrue* :: *Bool* $\rightarrow$ *Space* ()
> *isTrue* = *Factor* $\circ$ *indicator*

where *isTrue c* is the subspace which has measure 1 if c is true and 0 otherwise. The subspace of s where e holds is then the first projection of $\Sigma\ s$ (*isTrue* $\circ$ *e*).

> *subspace* :: ($a \rightarrow Bool$) $\rightarrow$ *Space a* $\rightarrow$ *Space a*
> *subspace e s* = *Project fst* ($\Sigma\ s$ (*isTrue* $\circ$ *e*))

We can show that if s has a non-zero measure, then the two definitions are equivalent:

Lemma 9.3. *measure s* $*$ *probability s e* = *measure* ($\Sigma\ s$ (*isTrue* $\circ$ *e*))

Proof. (where we shorten *indicator* to *ind* and *integrator s* to $\int \{s\}$)

> *measure_sigma_equations s e* =
>   *probability1 s e*
>   === {- Def. of *probability1* -}
>   *expectedValue s* (*ind* $\circ$ *e*)
>   === {- Def. of *expectedValue* -}
> $\int \{s\}$ (*ind* $\circ$ *e*) / *measure s*
>   === {- Def. of ($\circ$) -}

$\int \{s\} \ (\lambda x \rightarrow ind \ (e \ x)) \ / \ measure \ s$
=== {- mutiplication by 1 -}
$\int \{s\} \ (\lambda x \rightarrow ind \ (e \ x) * const \ 1 \ (x, ())) \ / \ measure \ s$
=== {- Def. of $\int \{\cdot\}$ -}
$\int \{s\} \ (\lambda x \rightarrow \int \{Factor \ (ind \ (e \ x))\} \ (\lambda y \rightarrow const \ 1 \ (x, y))) \ / \ measure \ s$
=== {- Def. of *isTrue* -}
$\int \{s\} \ (\lambda x \rightarrow \int \{isTrue \ (e \ x)\} \ (\lambda y \rightarrow const \ 1 \ (x, y))) \ / \ measure \ s$
=== {- Def. of $\int \{\cdot\}$ for Σ -}
$measure \ (\Sigma \ s \ (isTrue \circ e)) \ / \ measure \ s$

$\square$

We can now note that the space *observing e s* is the subspace of *s* where the event *e* is observed to be true — a kind of subspace which is often used in textbook problems.

It will often be convenient to define a space whose underlying set is a Boolean value and compute the probability of the identity event:

probability :: *Space Bool* $\rightarrow \mathbb{R}$
probability d = *expectedValue d indicator*

Sometimes one even finds in the literature and folklore the notation $P(v)$, where v is a value, which stands for $P(t = v)$, for an implicit random variable t. Here even more imagination is required from the reader, who must not only infer the space of outcomes, but also which random variable the author means.

9.6.1 Conditional probability

In Section 7.4.6, we encountered the notion of conditional probability, traditionally written $P(F \mid G)$ and read "probability of F given G". As suggested in Section 7.4.6 and brushed upon in Exercise 3.9, it is not the case that the expression $(F \mid G)$ is an event. Rather, a conditional probability is a separate concept, which takes both f and g as arguments. It is defined as the probability of f in the sub space where g holds:

condProb :: *Space a* $\rightarrow (a \rightarrow Bool) \rightarrow (a \rightarrow Bool) \rightarrow \mathbb{R}$
condProb s f g = *probability1* (*subspace g s*) *f*

We find the above definition more intuitive than the more usual definition $P(F \mid G) = P(F \cap G)/P(G)$. Why? Because it makes clear that, in $P(F \mid G)$, the event G acts as the subspace upon which the truth of F is integrated. (In fact, the $P(F \mid G)$ notation is an improvement over the $P(F)$ notation, in the sense that the underlying space is more explicit.)

Regardless, the equivalence between the two definitions can be proven, by symbolic calculation:

Lemma 9.4. *condProb s f g $==$ probability s $(\lambda y \to f\ y \wedge g\ y)$ / probability s g*

Proof.

$$
\begin{aligned}
&cond\_prob\_equations :: Space\ a \to (a \to Bool) \to (a \to Bool) \to \mathbb{R} \\
&cond\_prob\_equations\ s\ f\ g = \\
&\quad condProb\ s\ f\ g \\
&\quad === \{\text{- Def of condProb -}\} \\
&\quad probability1\ (subspace\ g\ s)\ f \\
&\quad === \{\text{- Def of subspace -}\} \\
&\quad probability1\ (\Sigma\ s\ (isTrue \circ g))\ (f \circ fst) \\
&\quad === \{\text{- Def of probability1 -}\} \\
&\quad expectedValue\ (\Sigma\ s\ (isTrue \circ g))\ (ind \circ f \circ fst) \\
&\quad === \{\text{- Def of expectedValue -}\} \\
&\quad (1\ /\ measure\ (\Sigma\ s\ (isTrue \circ g))) * (\int \{\Sigma\ s\ (isTrue \circ g)\}\ (ind \circ f \circ fst)) \\
&\quad === \{\text{- Def of } \int \{\cdot\}\ \text{(Sigma) -}\} \\
&\quad (1\ /\ measure\ s\ /\ probability1\ s\ g) * (\ \int \{s\} \qquad\quad \$\ \lambda x \to \\
&\qquad\qquad\qquad\qquad\qquad\qquad\qquad \int \{isTrue\ (g\ x)\}\ \$\ \lambda y \to \\
&\qquad\qquad\qquad\qquad\qquad\qquad\qquad ind \circ f \circ fst\ \$\ (x,y)) \\
&\quad === \{\text{- Def of } fst \text{ -}\} \\
&\quad (1\ /\ measure\ s\ /\ probability1\ s\ g) * (\ \int \{s\} \qquad\quad \$\ \lambda x \to \\
&\qquad\qquad\qquad\qquad\qquad\qquad\qquad \int \{isTrue\ (g\ x)\}\ \$\ \lambda y \to \\
&\qquad\qquad\qquad\qquad\qquad\qquad\qquad ind \circ f\ \$\ x) \\
&\quad === \{\text{- Def of } isTrue \text{ -}\} \\
&\quad (1\ /\ measure\ s\ /\ probability1\ s\ g) * (\int \{s\}\ \$\ \lambda x \to ind\ (g\ x) * ind\ (f\ x)) \\
&\quad === \{\text{- Property of } ind \text{ -}\} \\
&\quad (1\ /\ measure\ s\ /\ probability1\ s\ g) * (\int \{s\}\ \$\ \lambda x \to ind\ (g\ x \wedge f\ x)) \\
&\quad === \{\text{- associativity of multiplication -}\} \\
&\quad (1\ /\ probability1\ s\ g) * (\int \{s\}\ \$\ \lambda x \to ind\ (g\ x \wedge f\ x))\ /\ measure\ s \\
&\quad === \{\text{- Definition of } probability1 \text{ -}\} \\
&\quad (1\ /\ probability1\ s\ g) * probability1\ s\ (\lambda x \to g\ x \wedge f\ x)
\end{aligned}
$$

$\square$

9.6.2 Independent events

Another important notion in probability theory is that of independent events. One way to define independent events is as follows. E is independent from F iff $P(E \mid F) = P(E)$. According to Grinstead and Snell [2003], two events are independent iff. $P(E \cap F) = P(E) \cdot P(F)$. The proof that these formulations are equivalent can be written in the traditional notation as follows:

Proof. In the left to right direction:

$$P\ (E \cap F) \qquad = \{\text{- by def. of cond. prob -}\}$$
$$P\ (E \mid F) \cdot P\ (F) = \{\text{- by def. of independent events -}\}$$
$$P\ (E) \cdot P\ (F)$$

In the right to left direction:

$$P\ (E \mid F) \qquad\qquad = \{\text{- by def. of cond. prob -}\}$$
$$P\ (E \cap F)\ /\ P\ (F) \qquad = \{\text{- by assumption -}\}$$
$$P\ (E) \cdot P\ (F)\ /\ P\ (F) = \{\text{- by computation -}\}$$
$$P\ (E)$$

□

Let us now express the same definitions and the same theorem and proof in our DSL. The definition for independent events is:

> *independentEvents* :: *Space a* → (*a* → *Bool*) → (*a* → *Bool*) → *Bool*
> *independentEvents s e f* = *probability1 s e* == *condProb s e f*

The equivalent formulation is:

> *independentEvents2 s e f* =
> *probability1 s* (λx → *e x* $\wedge$ *f x*) == *probability1 s e* * *probability1 s f*

We can now state (and prove) the lemma using the DSL notation:

Lemma 9.5. independentEvents s e f $\Leftrightarrow$ independentEvents2 s e f

Proof. Left to right direction:

> *probability1 s* (λx → *e x* $\wedge$ *f x*)
> === {- by Lemma 9.4 -}
> *condProb s e f* * *probability1 s f*
> === {- by assumption -}
> *probability1 s e* * *probability1 s f*

□

We note that at this level of abstraction, the proofs follow the same structure as the textbook proofs — the underlying space s is constant.

Exercise 9.6. Express the rest of the proof using our DSL.

9.7   Examples: Dice, Drugs, Monty Hall

We are now ready to solve all three problems motivating this chapter.

9.7.1   Dice problem

```
diceSpace :: Space Bool
diceSpace =
      -- consider only the event "product >= 10"
    Project (λ(x, y) → (x * y ⩾ 10)) $
      -- observe that the sum is >= 7
    observing (λ(x, y) → (x + y ⩾ 7)) $
      -- sample two balanced die
    twoDice
```

Then we can compute its probability:

```
diceProblem :: ℝ
diceProblem = probability diceSpace

-- ⋙ diceProblem
-- 0.9047619047619047
```

Exercise 9.7. Use the monadic interface to define the same experiment.

To illustrate the use of the various combinators from above to explore a sample space we can compute a few partial results explaining the *diceProblem*:

```
p₁ (x, y) = x + y ⩾ 7
p₂ (x, y) = x * y ⩾ 10
test1     = measure (observing p₁ twoDice)                          -- 21
test2     = measure (observing p₂ twoDice)                          -- 19
testBoth  = measure (observing (λxy → p₁ xy ∧ p₂ xy) twoDice)      -- 19
prob21    = condProb (prod d6 d6) p₂ p₁                             -- 19/21
```

We can see that 21 possibilities give a sum $⩾ 7$, that 19 possibilities give a product $⩾ 10$ and that all of those 19 satisfy both requirements.

9.7.2 Drug test

The above drug test problem (Item 2. at the start of this chapter) is often used as an illustration for Bayes' theorem. We can solve it in exactly the same fashion as the Dice problem.

We begin by describing the space of situations. To do so we make heavy use of the *bernoulli* distribution. First we model the distribution of drug users.

> **type** *DrugUser* = *Bool*
> *drugUser* :: *Space Bool*
> *drugUser* = *bernoulli* 0.005

Then we model the distribution of test outcomes — depending on whether we have a user or not.

> **type** *TestResult* = *Bool*
> *testResult* :: *DrugUser* → *Space TestResult*
> *testResult isUser* = *bernoulli* (**if** *isUser* **then** 0.99 **else** 0.01)

Then we "zoom in on" (observe) only those with a positive test:

> *positiveTests* :: *Space* (*DrugUser*, *TestResult*)
> *positiveTests* = *observing snd* (Σ *drugUser testResult*)

Finally, because the test result is always *True*, we select only the *DrugUser* component.

> *drugSpace* :: *Space DrugUser*
> *drugSpace* = *Project fst positiveTests*

The probability is computed as usual:

> *userProb* :: ℝ
> *userProb* = *probability drugSpace*
>
> -- >>> *userProb*
> -- 0.33221476510067116

Thus a randomly selected individual with a positive test is a drug user with probability around one third (and about two thirds are false positives).

Perhaps surprisingly, we never needed Bayes' theorem to solve the problem. Indeed, Bayes' theorem is already incorporated in our definition of *probability*, so our methodology guarantees that we always respect it.

9.7.3   Monty Hall

We can model the Monty Hall problem as follows. For expository purposes, let us define the list of doors, and a uniform distribution of these doors.

type *Door* = *Int*
doors :: [*Door*]
doors = [1,2,3]
anyDoor :: *Distr Door*
anyDoor = *uniformDiscrete doors*

The event of "winning" depends on four variables:

- which door is the winning door (*winningDoor* :: *Door*)

- the initial door choice (*initiallyPickedDoor* :: *Door*)

- the door which Monty opens (*montyPickedDoor* :: *Door*)

- and finally, whether the player changes their mind after Monty shows that the door has a goat (*changing* :: *Bool*).

One way to do it is as follows:

haveWon :: *Bool* → ((*Door*, *Door*), *Door*) → *Bool*
haveWon changing ((*winningDoor*, *initiallyPickedDoor*), *montyPickedDoor*)
    = *finalChoice* == *winningDoor*
    where *finalChoice* = **case** *changing* **of**
    False → *initiallyPickedDoor*
    True → *head* (*doors* ∖ [*initiallyPickedDoor*, *montyPickedDoor*])

The player wins if their final choice is the right one. If they do not change their mind, then the final choice is equal to the initial one. If they do change their mind, then the final choice is neither their initial choice nor Monty's door. (Because there are only three doors there is only one door left.)

Then, we need to describe the set of situations, as a triple:

$$((winningDoor, initiallyPickedDoor), montyPickedDoor).$$

The first two variables are uniform, but the *montyPickedDoor* is uniform in the set (*doors* ∖ [*pickedDoor*, *winningDoor*]). Note that if the *initiallyPickedDoor* is the same as the *winningDoor*, then Monty has two choices. We then imagine that Monty picks a door at random among those, even though this might not be the case in a real game.[8]

---

[8]Exercise: check other strategies for Monty.

```
montySpace :: Space ((Door, Door), Door)
montySpace =
  (Σ (prod anyDoor anyDoor)
    (λ(winningDoor, pickedDoor) →
       uniformDiscrete (doors ∖ [pickedDoor, winningDoor])))
montyProblem :: Bool → Space Bool
montyProblem changing = Project (haveWon changing) montySpace

  -- >>> probability (montyProblem False)
  -- 0.3333333333333333

  -- >>> probability (montyProblem True)
  -- 0.6666666666666666
```

Thus, the "changing door" strategy is twice as good.

The Monty Hall is sometimes considered paradoxical: it is strange that changing one's mind can change the outcome. The crucial point to see this is that Monty will never reveal a door which contains the prize – that would end the show a bit too early. To illustrate, an *incorrect* way to model the Monty Hall space of situations is the following:

```
montySpaceIncorrect :: Space ((Door, Door), Door)
montySpaceIncorrect =
  (Σ (prod anyDoor anyDoor)
    (λ(_, pickedDoor) → uniformDiscrete (doors ∖ [pickedDoor])))
  -- >>> probability (Project (haveWon False) montySpaceIncorrect)
  -- 0.5
```

The above does not correctly model the problem, because it allows Monty to reveal the winning door, while the problem specification said he would reveal a goat.

9.8 *Solving a problem with equational reasoning

Consider the following problem: how many times must one throw a coin before one obtains 3 heads in a row? We can model the problem as follows:

```
coin :: Distr Bool
coin = bernoulli 0.5

coins :: Distr [Bool]
coins = fmap (λ(x, xs) → x : xs)
             (prod coin coins)
```

$threeHeads :: [Bool] \rightarrow Int$
$threeHeads\ (True:True:True:\_) = 3$
$threeHeads\ (\_:xs) = 1 + threeHeads\ xs$

$example' :: Space\ Int$
$example' = fmap\ threeHeads\ coins$

Even though the problem is easy to *model* using the DSL, it is not easy to compute a solution. Indeed, attempting to evaluate *probability1 threeHeads* (<5) does not terminate. This is because we have an infinite list, which translates to infinitely many cases to consider. So the evaluator cannot solve this problem in finite time. Hence, we have to resort to a symbolic method to solve it. This will require extensive symbolic reasoning (perhaps not for the faint of heart). One may skip all the equational reasoning in first reading: the important point is to realise that one is able to write down the kind of proofs that one would do with pen and paper directly within a program. (We even use the Haskell type-checker to verify that each line typechecks. Even though this does not guarantee that the proof is sound, we catch most typos this way.)

The first step in our computation of the solution is a creative one, which involves generalising the problem to computing the number of throws to obtain the remaining *m* heads in a row, where *m* is between zero and three. For this purpose we define the function *tH*, such that *tH* 3 == *threeHeads*:

$tH :: Int \rightarrow [Bool] \rightarrow Int$
$tH\ 0\ \_ = 0$ -- no heads remaining means 0 more heads are needed
$tH\ m\ (x:xs) = 1 +$ -- if $m > 0$ heads remain, we need one throw +
$\qquad\qquad \textbf{if}\ x$ -- if this throw is a head
$\qquad\qquad \textbf{then}\ tH\ (m-1)\ xs$ -- we cont. needing $m-1$ heads
$\qquad\qquad \textbf{else}\ tH\ 3\ xs$ -- otherwise we start all over

We then define *helper m = fmap (tH m) coins* which returns a *Distr Int* representing the probability distribution of throws remaining. Computing it symbolically will give our answer. But first, we need a lemma which helps us push *fmap* inside Σ:

Lemma 9.6. $fmap\ f\ (\Sigma\ a\ g)$ == $fmap\ snd\ (\Sigma\ a\ (\lambda x \rightarrow fmap\ (f \circ (x,))\ (g\ x)))$

Proof. We check the equivalence by using semantic equality:

$project\_sigma\_equations\ f\ a\ g\ h =$
$\quad \int \{fmap\ f\ (\Sigma\ a\ g)\}\ h$
$\quad === \quad$ -- by def
$\quad \int \{\Sigma\ a\ g\}\ (h \circ f)$
$\quad === \quad$ -- by def

$$\int \{a\}\ (\lambda x \to \int \{g\ x\}\ (\lambda y \to (h \circ f)\ (x,y)))$$
=== -- rewriting in point-free style
$$\int \{a\}\ (\lambda x \to \int \{g\ x\}\ (h \circ f \circ (x,)))$$
=== -- by def of integrator of *fmap*
$$\int \{a\}\ (\lambda x \to \int \{fmap\ (f \circ (x,))\ (g\ x)\}\ h)$$
=== -- $h == \lambda y \to h\ y$
$$\int \{a\}\ (\lambda x \to \int \{fmap\ (f \circ (x,))\ (g\ x)\}\ \$\ \lambda y \to h\ y)$$
=== -- taking an explicit (x,y) pair
$$\int \{a\}\ (\lambda x \to \int \{fmap\ (f \circ (x,))\ (g\ x)\}\ \$\ \lambda y \to (h \circ snd)\ (x,y))$$
=== -- by def of integrator of Σ
$$\int \{\Sigma\ a\ (\lambda x \to fmap\ (f \circ (x,))\ (g\ x))\}\ (h \circ snd)$$
=== -- by def of integrator of *fmap*
$$\int \{fmap\ snd\ (\Sigma\ a\ (\lambda x \to fmap\ (f \circ (x,))\ (g\ x)))\}\ h$$

□

Lemma 9.7. *fmap* (**if** c **then** a **else** b) $s ==$ **if** c **then** *fmap* a s **else** *fmap* b s

Proof. We check the equivalence by using semantic equality:

$$project\_if\_equations\ c\ a\ b\ s\ g =$$
$$\int \{fmap\ (\textbf{if}\ c\ \textbf{then}\ a\ \textbf{else}\ b)\ s\}\ g\ ===\ \text{-- by def}$$
$$\int \{s\}\ ((\textbf{if}\ c\ \textbf{then}\ a\ \textbf{else}\ b) \circ g)\ ===\ \text{-- by def}$$
$$\int \{\ \textbf{if}\ c\ \textbf{then}\ fmap\ a\ s\ \textbf{else}\ fmap\ b\ s\}\ g$$

□

We can proceed from *helper m* by equational reasoning, and see what we get:

$$unfolding\_helper\_equations\ m =$$
$$\quad helper\ m$$
=== -- by def
$$\quad fmap\ (tH\ m)\ coins$$
=== -- by def
$$\quad fmap\ (tH\ m)\ (fmap\ (\lambda(x,xs) \to x:xs)\ (prod\ coin\ coins))$$
=== -- property of *fmap* (Exercise 9.1, functoriality of *Project*)
$$\quad fmap\ (tH\ m \circ \lambda(x,xs) \to x:xs)\ (prod\ coin\ coins)$$
=== -- by def
$$\quad fmap\ (\lambda(x,xs) \to 1 + \textbf{if}\ x\ \textbf{then}\ tH\ (m-1)\ xs\ \textbf{else}\ tH\ 3\ xs)$$
$$\quad\quad (prod\ coin\ coins)$$
=== -- by functoriality of *fmap*
$$\quad fmap\ (1+)\ (fmap\ (\lambda(x,xs) \to \textbf{if}\ x\ \textbf{then}\ tH\ (m-1)\ xs$$
$$\quad\quad\quad\quad\quad\quad\quad\quad\quad\quad \textbf{else}\ tH\ 3\ xs)$$
$$\quad\quad\quad (prod\ coin\ coins))$$

===   -- by Lemma 9.6
$fmap$ $(1+)$ $(fmap\ snd$
$$(\Sigma\ coin\ (\lambda x \to fmap\ (\lambda xs \to \textbf{if}\ x\ \textbf{then}\ tH\ (m-1)\ xs$$
$$\textbf{else}\ \ tH\ 3\ xs)$$
$$coins)))$$

===   -- by functoriality of *fmap*
$fmap$ $((1+) \circ snd)$
$$(\Sigma\ coin\ (\lambda x \to fmap\ (\lambda xs \to \textbf{if}\ x\ \textbf{then}\ tH\ (m-1)\ xs$$
$$\textbf{else}\ \ tH\ 3\ xs)$$
$$coins))$$

===   -- by semantics of **if** in Haskell
$fmap$ $((1+) \circ snd)$
$$(\Sigma\ coin\ (\lambda x \to fmap\ (\textbf{if}\ x\ \textbf{then}\ (tH\ (m-1))$$
$$\textbf{else}\ \ (tH\ 3))$$
$$coins))$$

===   -- by semantics of **if** in Haskell
$fmap$ $((1+) \circ snd)$
$$(\Sigma\ coin\ (\lambda x \to \textbf{if}\ x\ \ \textbf{then}\ fmap\ (tH\ (m-1))\ coins$$
$$\textbf{else}\ \ fmap\ (tH\ 3) \qquad coins))$$

===   -- by definition of *helper*.
$fmap$ $((1+) \circ snd)\ (\Sigma\ coin\ (\lambda x \to \textbf{if}\ x\ \textbf{then}\ helper\ (m-1)$
$$\textbf{else}\ \ helper\ 3))$$

To sum up:

$helper\ 0\ = point\ 0$
$helper\ m = fmap\ ((1+) \circ snd)$
$$(\Sigma\ coin\ (\lambda h \to$$
$$\textbf{if}\ h\ \textbf{then}\ helper\ (m-1)$$
$$\textbf{else}\ \ helper\ 3))\ \ \ \text{-- when we have a tail, we restart}$$

Evaluating the probability still does not terminate: we no longer have an infinite list, but we still have infinitely many possibilities to consider: however small, there is always a probability to get a "tail" at the wrong moment, and the evaluation must continue.

But we can keep performing our symbolic calculation. We can start by showing that *helper m* is a distribution (its measure is 1). The proof is by induction on m:

- for the base case *measure (helper 0) = measure (pure 0) = 1*

- for the step case: assume that *helper m* is a distribution. Then, for every h, **if** h **then** *helper m* **else** *helper 3* is a distribution too. The result is obtained by using the distribution property of Σ (Lemma 9.2).

Then, we can symbolically compute the integrator of *helper*. The base case is $\int \{helper\ 0\}\ id\ ==\ 0$, and left as an exercise to the reader. In the recursive case, we have:

$\int \{helper\ (m+1)\}\ id$
== {- By above result regarding *helper* -}
$\int \{fmap\ ((1+)\circ snd)$
    $(\Sigma\ coin\ (\lambda h \rightarrow$ **if** h **then** $helper\ (m-1)$ **else** $helper\ 3))\}\ id$
== {- By integrator def -}
$\int \{\Sigma\ coin\ (\lambda h \rightarrow$ **if** h **then** $helper\ (m-1)$ **else** $helper\ 3)\}\ ((1+)\circ snd)$
== {- By integrator def -}
$\int \{coin\}\ \$\ \lambda h \rightarrow \int \{$ **if** h **then** $helper\ m$ **else** $helper\ 3\}\ (1+)$
== {- By linearity of $\int \{s\}$ -}
$1 + \int \{coin\}\ \$\ \lambda h \rightarrow \int \{$ **if** h **then** $helper\ m$ **else** $helper\ 3\}\ id$
== {- By case analysis -}
$1 + (\int \{coin\}\ \$\ \lambda h \rightarrow$ **if** h **then** $\int \{helper\ m\}\ id$ **else** $\int \{helper\ 3\}\ id)$
== {- By integrator/bernoulli (Exercise 9.3) -}
$1 + 0.5 * \int \{helper\ m\}\ id + 0.5 * \int \{helper\ 3\}\ id$

If we let $h\ m\ =\ expectedValueOfDistr\ (helper\ m)$, and using the above lemma, then we find:

$h\ (m+1)\ ==\ 1 + 0.5 * h\ m + 0.5 * h\ 3$

which we can rewrite as:

$2 * h\ (m+1) = 2 + h\ m + h\ 3$

and expand for $m = 0$ to $m = 2$:

$2 * h\ 1 = 2 + 0 +\quad h\ 3$
$2 * h\ 2 = 2 + h\ 1 + h\ 3$
$2 * h\ 3 = 2 + h\ 2 + h\ 3$

This leaves us with a system of linear equations with three unknowns $h\ 1, h\ 2$ and $h\ 3$. Eliminating $h\ 3 = 2 * h\ 1 - 2$ and simplifying gives

$3 * h\ 1 = 2 * h\ 2$
$2 * h\ 1 = 4 + h\ 2$

and eliminating $h\ 2 = 2 * h\ 1 - 4 +$ simplifying gives

$h\ 1 = 8; h\ 2 = 12; h\ 3 = 14$

Thus the answer to the initial problem is: we can expect to need 14 coin flips to get three heads in a row.

Appendix A

The course "DSLs of Mathematics"

From 2016 there has been a BSc-level university course on "Domain-Specific Languages of Mathematics (DSLM)" at the Computer Science and Engineering (CSE) Department, joint between Chalmers University of Technology and University of Gothenburg, Sweden. The learning outcomes of the course are presented in Figure A.1.

In the first instances, the course is an elective course for the second year within programmes such as CSE[1], SE, and Math. The potential students will have all taken first-year mathematics courses, and the only prerequisite which some of them will not satisfy will be familiarity with functional programming. However, as some of the current data structures course (common to the Math and CSE programmes) shows, math students are usually able to catch up fairly quickly, and in any case we aim to keep to a restricted subset of Haskell (no "advanced" features are required).

The formal course prerequisites say that the student should have successfully completed

- a course in discrete mathematics as for example Introductory Discrete Mathematics.

- 15 hec in mathematics, for example Linear Algebra and Calculus

---

[1]CSE = Computer Science & Engineering = Datateknik = D

247

- 15 hec in computer science, for example (Introduction to Programming or Programming with MATLAB) and Object-oriented Software Development

- an additional 22.5 hec of any mathematics or computer science courses.

The unit here is "higher education credit (hec)" where 60 hec corresponds to one full-time year of study.

To assess the impact in terms of increased quality of education, we planned to measure how well the students do in ulterior courses that require mathematical competence (in the case of engineering students) or software competence (in the case of math students). For math students, we would like to measure their performance in ulterior scientific computing courses, but we have taught too few math students so far to make good statistics. But for CSE students we have measured the percentage of students who, having taken DSLM, pass the third-year courses *Transforms, signals and systems (TSS)* and *Control Theory (sv: Reglerteknik)*, which are current major stumbling blocks. We have compared the results with those of a control group (students who have not taken the course). The evaluation of the student results shows improvements in the pass rates and grades in later courses. This is very briefly summarised in Table A.1 and more details are explained by Jansson et al. [2018].

- Knowledge and understanding
  - design and implement a DSL for a new domain
  - organize areas of mathematics in DSL terms
  - explain main concepts of elementary real and complex analysis, algebra, and linear algebra

- Skills and abilities
  - develop adequate notation for mathematical concepts
  - perform calculational proofs
  - use power series for solving differential equations
  - use Laplace transforms for solving differential equations

- Judgement and approach
  - discuss and compare different software implementations of mathematical concepts

Figure A.1: Learning outcomes

The work that leads up to the current book started in 2014 with an assessment of what prerequisites we can reasonably assume and what mathematical fields the targeted students are likely to encounter in later studies. In 2015 we submitted a course plan so that the first instance of the course could start early 2016. We also surveyed similar courses being offered at other universities, but did not find any close matches. ("The Haskell road to Logic, Maths and

	PASS	IN	OUT
TSS pass rate	77%	57%	36%
TSS mean grade	4.23	4.10	3.58
Control pass rate	68%	45%	40%
Control mean grade	3.91	3.88	3.35

Table A.1: Pass rate and mean grade in third year courses for students who took and passed DSLsofMath and those who did not. Group sizes: PASS 34, IN 53, OUT 92 (145 in all).

Programming" by Doets and van Eijck [2004] is perhaps the closest, but it is mainly aimed at discrete mathematics.)

While preparing course materials for use within the first instance we wrote a paper [Ionescu and Jansson, 2016] about the course and presented the pedagogical ideas at several events (TFPIE'15, DSLDI'15, IFIP WG 2.1 #73 in Göteborg, LiVe4CS in Glasgow). In the following years we used the feedback from students following the standard course evaluation in order to improve and further develop the course material into complete lecture notes, and now a book.

In the first few years, the enrolment and results of the DSLsofMath course itself was as follows:

Year	'16	'17	'18	'19	'20	'21
Student count	28	43	39	59	50	67
Pass rate (%)	68	58	89	73	68	72

Note that this also counts students from other programmes (mainly SE and Math) while Table A.1 only deals with the CSE programme students.

Appendix B

Parameterised Complex Numbers

module *DSLsofMath.CSem* (**module** *DSLsofMath.CSem,*
   module *DSLsofMath.Algebra*) **where**
import *Prelude hiding* (*Num* (..), (/), (^), *Fractional* (..), *Floating* (..), *sum*)
import *DSLsofMath.Algebra*
   (*Additive* (*zero*, (+)),   *AddGroup* (*negate*), (−), *Ring*
   , *Multiplicative* (*one*, (*)), (^+), *MulGroup* ((/)),   *Field*
)

newtype *Complex r* = *C* (*r*, *r*) **deriving** *Eq*

Lifting operations to a parameterised type   When we define addition on complex numbers (represented as pairs of real and imaginary components) we can do that for any underlying type *r* which supports addition. Note that *liftCS* takes (+) as its first parameter and uses it twice on the RHS.

type *CS* = *Complex*   -- for shorter type expressions below
liftCS :: (*r* →   *r* →   *r*) →
           (*CS*   *r* → *CS r* → *CS r*)
liftCS (+) (*C* (*x*, *y*)) (*C* (*x'*, *y'*)) = *C* (*x* + *x'*, *y* + *y'*)
addC :: *Additive r* ⇒ *Complex r* → *Complex r* → *Complex r*
addC = *liftCS* (+)
toC :: *Additive r* ⇒ *r* → *Complex r*
toC x = *C* (*x*, *zero*)

251

mulC :: *Ring r* ⇒   *Complex r* → *Complex r* → *Complex r*
mulC (*C* (*ar*, *a_i*)) (*C* (*br*, *b_i*)) = *C* (*ar* ∗ *br* − *a_i* ∗ *b_i*, *ar* ∗ *b_i* + *a_i* ∗ *br*)
modulusSquaredC :: *Ring r* ⇒ *Complex r* → *r*
modulusSquaredC (*C* (*x*, *y*)) = *x*^+2 + *y*^+2
scaleC :: *Multiplicative r* ⇒ *r* → *Complex r* → *Complex r*
scaleC a (*C* (*x*, *y*)) = *C* (*a* ∗ *x*, *a* ∗ *y*)
conj :: *AddGroup r* ⇒ *Complex r* → *Complex r*
conj (*C* (*x*, *y*)) = *C* (*x*, *negate y*)

instance *Additive r* ⇒ *Additive* (*Complex r*) **where**
    (+) = *addC*
    zero = *toC zero*

instance *AddGroup r* ⇒ *AddGroup* (*Complex r*) **where**
    negate (*C* (*a*, *b*)) = *C* (*negate a*, *negate b*)

instance *Ring r* ⇒ *Multiplicative* (*Complex r*) **where**
    (∗) = *mulC*
    one = *toC one*

instance *Field r* ⇒ *MulGroup* (*Complex r*) **where**
    (/) = *divC*

divC :: *Field a* ⇒ *Complex a* → *Complex a* → *Complex a*
divC x y = *scaleC* (*one* / *modSq*) (*x* ∗ *conj y*)
    where *modSq* = *modulusSquaredC y*

re :: *Complex r* → *r*
re z@(*C* (*x*, *y*)) = *x*
im :: *Complex r* → *r*
im z@(*C* (*x*, *y*)) = *y*

instance *Show r* ⇒ *Show* (*Complex r*) **where**
    show = *showCS*

showCS :: *Show r* ⇒ *Complex r* → *String*
showCS (*C* (*x*, *y*)) = *show x* ++ " + " ++ *show y* ++ "i"

A corresponding syntax type: the second parameter *r* makes is possible to express complex numbers over different base types (like *Double*, *Float*, $\mathbb{Z}$, etc.).

data *ComplexSy v r* = *Var v*
                 | *FromCart r r*
                 | *ComplexSy v r* :++ *ComplexSy v r*
                 | *ComplexSy v r* :∗∗ *ComplexSy v r*

Bibliography

R. A. Adams and C. Essex. *Calculus: a complete course*. Pearson Canada, 7th edition, 2010.

J. Bernardy, P. Jansson, and K. Claessen. Testing polymorphic properties. In A. D. Gordon, editor, *ESOP 2010*, volume 6012 of *LNCS*, pages 125–144. Springer, 2010. doi:10.1007/978-3-642-11957-6_8.

J. Bernardy, S. Chatzikyriakidis, and A. Maskharashvili. A computational treatment of anaphora and its algorithmic implementation. *Journal of Logic, Language and Information*, 30(1):1–29, 2021. doi:10.1007/s10849-020-09322-7.

R. Bird and P. Wadler. *Introduction to Functional Programming, 1988*. Prentice-Hall, Englewood Cliffs, NJ, 1988.

N. Botta, P. Jansson, and C. Ionescu. Contributions to a computational theory of policy advice and avoidability. *J. Funct. Program.*, 27:e23, 2017a. ISSN 0956-7968. doi:10.1017/S0956796817000156.

N. Botta, P. Jansson, C. Ionescu, D. R. Christiansen, and E. Brady. Sequential decision problems, dependent types and generic solutions. *Logical Methods in Computer Science*, 13(1), 2017b. doi:10.23638/LMCS-13(1:7)2017.

N. Botta, P. Jansson, and C. Ionescu. The impact of uncertainty on optimal emission policies. *Earth System Dynamics*, 9(2):525–542, 2018. doi:10.5194/esd-9-525-2018.

R. Boute. The decibel done right: a matter of engineering the math. *Antennas and Propagation Magazine, IEEE*, 51(6):177–184, 2009. doi:10.1109/MAP.2009.5433137.

E. Brady. *Type-driven Development With Idris*. Manning, 2016. ISBN 9781617293023. URL http://www.worldcat.org/isbn/9781617293023.

K. Claessen and J. Hughes. QuickCheck: a lightweight tool for random testing of Haskell programs. In *Proc. ACM SIGPLAN international conference on Funct. Prog.*, pages 268–279. ACM, 2000. doi:10.1145/1988042.1988046.

L. M. de Moura, S. Kong, J. Avigad, F. van Doorn, and J. von Raumer. The lean theorem prover (system description). In A. P. Felty and A. Middeldorp, editors, *CADE*, volume 9195 of *LNCS*, pages 378–388. Springer, 2015. doi:10.1007/978-3-319-21401-6_26.

K. Doets and J. van Eijck. *The Haskell Road to Logic, Maths and Programming*. Texts in computing. King's College Publications, London, 2004. ISBN 978-0-9543006-9-2. URL https://fldit-www.cs.uni-dortmund.de/~peter/PS07/HR.pdf.

J. Duregård, P. Jansson, and M. Wang. Feat: functional enumeration of algebraic types. In J. Voigtländer, editor, *ACM SIGPLAN Symposium on Haskell*, pages 61–72. ACM, 2012. doi:10.1145/2364506.2364515.

C. H. Edwards, D. E. Penney, and D. Calvis. *Elementary Differential Equations*. Pearson Prentice Hall Upper Saddle River, NJ, 6h edition, 2008.

D. Gries and F. B. Schneider. *A logical approach to discrete math*. Springer, 1993. doi:10.1007/978-1-4757-3837-7.

D. Gries and F. B. Schneider. Teaching math more effectively, through calculational proofs. *American Mathematical Monthly*, pages 691–697, 1995. doi:10.2307/2974638.

C. M. Grinstead and J. L. Snell. *Introduction to Probability*. AMS, 2003. URL http://www.dartmouth.edu/~chance/teaching_aids/books_articles/probability_book/book.html.

R. Hinze and A. Löh. *Guide to lhs2TeX*, 2020. The tool lhs2tex is a preprocessor for typesetting Haskell sources with LaTeX. Available from https://hackage.haskell.org/package/lhs2tex-1.24.

C. Ionescu and P. Jansson. Testing versus proving in climate impact research. In N. A. Danielsson and B. Nordström, editors, *Proc. Workshop Types for Proofs and Programs (TYPES'11)*, volume 19 of *LIPIcs*, pages 41–54, 2011. doi:10.4230/LIPIcs.TYPES.2011.41.

C. Ionescu and P. Jansson. Dependently-typed programming in scientific computing - examples from economic modelling. In R. Hinze, editor, *Implementation and Application of Functional Languages*, volume 8241 of *LNCS*, pages 140–156. Springer, 2012. doi:10.1007/978-3-642-41582-1_9.

C. Ionescu and P. Jansson. Domain-specific languages of mathematics: Presenting mathematical analysis using functional programming. In J. Jeuring and J. McCarthy, editors, *Proc. International Workshop on Trends in Functional Programming in Education*, volume 230 of *EPTCS*, pages 1–15, 2016. doi:10.4204/EPTCS.230.1.

P. Jansson, S. H. Einarsdóttir, and C. Ionescu. Examples and results from a BSc-level course on domain specific languages of mathematics. In P. Achten and H. Miller, editors, *Proc. 7th Int. Workshop on Trends in Functional Programming in Education*, volume 295 of *EPTCS*, pages 79–90, 2018. doi:10.4204/EPTCS.295.6. Presented at TFPIE 2018.

P. Jansson, C. Ionescu, and J.-P. Bernardy. *Domain-Specific Languages of Mathematics*, volume 24 of *Texts in Computing*. College Publications, 2022. ISBN 978-1-84890-388-3.

J. Jeuring, P. Jansson, and C. Amaral. Testing type class laws. In J. Voigtländer, editor, *Proceedings of the 2012 Haskell Symposium*, pages 49–60. ACM, 2012. doi:10.1145/2364506.2364514.

R. Kraft. Functions and parameterizations as objects to think with. In *Maple Summer Workshop, July 2004, Wilfrid Laurier University, Waterloo, Ontario, Canada*, 2004.

E. Landau. *Einführung in die Differentialrechnung und Integralrechnung*. Noordhoff, 1934.

E. Landau. *Differential and Integral Calculus*. AMS/Chelsea Publication Series. AMS Chelsea Pub., 2001.

D. Lincke, P. Jansson, M. Zalewski, and C. Ionescu. Generic libraries in C++ with concepts from high-level domain descriptions in Haskell. In W. M. Taha, editor, *IFIP Working Conf. on Domain-Specific Languages*, volume 5658 of *LNCS*, pages 236–261. Springer, 2009. doi:10.1007/978-3-642-03034-5_12.

S. Mac Lane. *Mathematics Form and function*. Springer New York, 1986.

M. D. McIlroy. Functional pearl: Power series, power serious. *J. of Functional Programming*, 9:323–335, 1999. doi:10.1017/S0956796899003299.

S. Mu, H. Ko, and P. Jansson. Algebra of programming using dependent types. In P. Audebaud and C. Paulin-Mohring, editors, *International Conference on Mathematics of Program Construction*, volume 5133 of *LNCS*, pages 268–283. Springer, 2008. doi:10.1007/978-3-540-70594-9_15.

S. Mu, H. Ko, and P. Jansson. Algebra of programming in Agda: Dependent types for relational program derivation. *J. Funct. Program.*, 19(5):545–579, 2009. doi:10.1017/S0956796809007345.

U. Norell. Dependently typed programming in Agda. In K. P., P. R., and S. D., editors, *Advanced Functional Programming*, volume 5832 of *LNCS*, pages 230–266. Springer, 2009. doi:10.1007/978-3-642-04652-0_5.

T. J. Quinn and S. Rai. Discovering the laplace transform in undergraduate differential equations. *PRIMUS*, 18(4):309–324, 2008. doi:10.1080/10511970601131613.

J. J. Rotman. *A first course in abstract algebra*. Pearson Prentice Hall, 2006.

W. Rudin. *Principles of mathematical analysis*, volume 3. McGraw-Hill New York, 1964.

W. Rudin. *Real and complex analysis*. Tata McGraw-Hill Education, 1987.

D. Stirzaker. *Elementary Probability*. Cambridge University Press, 2 edition, 2003. doi:10.1017/CBO9780511755309.

G. J. Sussman and J. Wisdom. *Functional Differential Geometry*. MIT Press, 2013.

J. Tolvanen. Industrial experiences on using DSLs in embedded software development. In *Proceedings of Embedded Software Engineering Kongress (Tagungsband), December 2011*, 2011. doi:10.1.1.700.1924.

C. Wells. Communicating mathematics: Useful ideas from computer science. *American Mathematical Monthly*, pages 397–408, 1995. doi:10.2307/2975030.

Index